Astronomers' Universe

More information about this series at http://www.springer.com/series/6960

Peter Linde

The Hunt for Alien Life

A Wider Perspective

 Springer

Peter Linde
Lund, Sweden

Original Swedish edition published by Karavan Förlag, Lund, 2013

ISSN 1614-659X   ISSN 2197-6651   (electronic)
Astronomers' Universe
ISBN 978-3-319-24116-6   ISBN 978-3-319-24118-0   (eBook)
DOI 10.1007/978-3-319-24118-0

Library of Congress Control Number: 2015956604

Springer Cham Heidelberg New York Dordrecht London
© Springer International Publishing Switzerland 2016

A laser, fired by one of the Unit Telescopes of the VLT, is pointing at the heart of the Milky Way.
Credit: G. Tremblay/ESO

Printed on acid-free paper

Springer International Publishing AG Switzerland is part of Springer Science+Business Media
(www.springer.com)

To Marcus

Preface

Everyone should have the opportunity of looking at a really dark, starry sky. In today's world, more than half of humanity lives in cities; in many countries it is eight out of ten. Observing the night side of the Earth from space reveals how the city illumination spreads more and more across the planet. The risk is increasing that many will never see what a real starry sky looks like. And yet, it is not so difficult. You only need to be a little lucky with the weather and leave the city during an autumn evening. I don't think I have ever met anyone that has stayed completely untouched by such a view. There is so much depth to it, not just physically but also intellectually, aesthetically and emotionally. The night sky provides a healthy reminder of the perspectives of life. "How small and insignificant we all are!" is a common reaction. I disagree. Instead, I get a feeling of affinity with the infinite out there, which steadily sends its light to the Earth.

Fascinatingly enough, modern science, to a large extent, confirms the close connection between ourselves and the universe. The atoms that build up our bodies were created out there, both at the time of the creation of the universe and through later processing deep inside many generations of stars. This is certainly also true of our brains. It is no exaggeration to claim that the brain is also a result of the universe, guided by Darwin's theory of evolution. This kilo and a half of brain substance must be regarded as the most highly developed and organised form of matter that we know of. In this respect, we are not small, on the contrary. At least here on Earth, matter has developed intelligence and sentience, a truly remarkable fact. At least here on Earth, the universe has made it possible to begin a study of itself. As a newborn infant, we have the opportunity to try to understand our universe. I believe it is our duty to accept that challenge, and my book reflects this.

Clearly, while looking at the starry sky, the idea is not far-fetched that somewhere other intelligent beings may exist. They might be watching a starry sky in the same way, probably unknown to us. Do they have the same feelings? Are they asking the same questions? Until now, the very thought of getting those answers seemed meaningless; how would we ever be able to find out?

In 1995, these questions became considerably less meaningless. For the first time, astronomers confirmed that there are planets not just in our own solar system but also orbiting other stars. The discovery was the beginning of an avalanche of development. Within only 20 years time, we have discovered around 2000 new planets. Soon, they got their own label—exoplanets. The discovery also became the beginning of a new, expanding, scientific discipline—astrobiology. So, now that we know that planets are common and that there may be hundreds of millions in just our own galaxy, the Milky Way, it becomes inevitable to again ask the question: is there life out there?

The ambition of this book is to give a background, in popular terms, to the hunt for alien life. It is a fascinating research to follow, as almost daily new exoplanets are reported. Other scientists devote their entire lives to try to catch intelligent messages, which they believe are currently under way toward Earth. And humanity itself is sending information out that may, one day, be collected by somebody else.

I have tried to anchor this review in the facts of modern research. At the same time, I have given space to some speculations about aliens and the future destiny of man. Some of these concepts border on the domain of science fiction, but I believe this can be allowed, as long as it remains in the realm of natural science. Hopefully, the reader can occasionally feel the "sense of wonder" that I myself often feel.

Writing this book has been a challenge in many ways. The research concerning exoplanets is developing rapidly, and it is unavoidable that facts given in the book will become obsolete with time. I have created a special Web site, www.peterlinde.net, where the latest information is available, along with a lot of background material. I have also made a point of trying to avoid too many technical terms. In cases where this has not been possible, I refer to the glossary of terms at the end of the book.

The subject covers many disciplines, and it is impossible to be an expert on all of them. I have therefore been helped by friends and experts which I want to thank. Ulf R Johansson, journalist and eternal astronomy enthusiast, gave me many tips and much encouragement. The microbiologist Leif Petersson provided valuable input on the chapter about early life on Earth. Anders Johansen, exoplanet researcher at Lund Observatory, made several improvements to the chapters about planets and exoplanets. The astronomer Björn Stenholm did the same, especially for the chapters on SETI and Drake's formula. Regarding the English edition, I want to thank the people at Springer, in particular Mike Carroll who used his magic wand to convert my English manuscript into a much more readable text. If, in spite of their efforts, there still remain errors in the text, it is entirely due to myself.

Finally, I would like to thank the two people closest to me, Eva Dagnegård, who follows me in life and has continually encouraged me, and my globetrotting daughter Amanda, who, when she turns 90 years in 2080, will have experienced a world very different from the current one, hopefully an even better one.

Lund, Sweden   Peter Linde

Contents

1. A Cosmic Perspective

The enormous distances of the universe have always posed a challenge for the human intellect. Things have not become easier by the fact that the universe has been continuously growing—both in a physical sense and in our conceptualisation of it. Our perception of the universe has developed over many centuries, but over the past century it has definitely undergone a revolution.

In parallel with the increasing insights about the real structure and size of the universe, the role and place for mankind has gradually needed modification. The Earth is no longer located at the centre of the universe, nor is mankind. Perhaps it is time—not only in terms of knowledge but also philosophically and morally—to accept this as a fact? And to entertain the prospect that there is life somewhere beyond Earth? The time is right to take a closer look at this possibility.

In this book I will try to give an outline of the universe of new possibilities. The scientists now know of almost 2000 planets that belong to other stars than our own. They estimate that the number of planets in our galaxy alone can be counted in many hundreds of millions. Among other things we will investigate how this came about. Could there be other intelligent civilisations out there? More than fifty years of listening to the universe so far have not led to any positive answers, but we will nevertheless dare to discuss the possibilities of getting in touch with hypothetical extraterrestrial intelligences. Might we even go and visit? To speculate is allowed, but is there any reality behind such dreams?

Let us begin with acquiring an appreciation of how big the universe really is. We are talking about enormous distances that need to be traversed in order to make possible any form of observation, communication or contact.

© Springer International Publishing Switzerland 2016
P. Linde, *The Hunt for Alien Life*, Astronomers' Universe,
DOI 10.1007/978-3-319-24118-0_1

An Expanding Universe

Observers in antiquity made realistic calculations of the size of the Earth, the Moon and the Sun. The distance to the Moon could also be estimated with reasonable precision, with the help, for example, of lunar eclipses, while the distance to the Sun was grossly underestimated. While not much scientific progress was made in the medieval ages, there were always people who wanted to know more, as illustrated in Fig. 1.1. The first telescopes used in the seventeenth century resulted in radical improvements of observational accuracy. The distance to the Sun, and thus the scale of our entire solar system, was now given more correct proportions. Among other things, astronomers made use of solar transits of the

FIG. 1.1 In a woodcut, probably made in the 1880s for a book by the French astronomy populariser Camille Flammarion, is illustrated medieval mans desire to break boundaries and find new knowledge

planet Venus. In the eighteenth century a large number of scientists ventured out on long journeys to observe this rare phenomenon from various places in the world. They arrived quite close to the correct value of the distance, which nowadays is determined to just over 149 million kilometres. This distance has become a special unit of length in astronomy known as an astronomical unit (AU).

But how far was it to the stars? That stars were very distant suns was always considered as a reasonable concept. Evidently, the distance had to be very long since the stars in our sky are so much fainter than the Sun.

Tycho Brahe was among the first to make a serious effort to measure these distances. The method he considered was based on the fact that a nearby object seems to move with respect to distant objects if you observe it from two different directions. This can easily be demonstrated by using the left eye to watch a nearby object and notice its position relative to its background. By changing to the right eye, the nearby object will have moved relative to the background, for instance a wall. We call this angular shift the parallax of the object. The idea was that stars would also show a parallax if precise measurements were made from different positions in the Earth's orbit around the Sun (Fig. 1.2). But although Tycho was thinking correctly and had access to the by far best angular measurement instruments of his time—although no telescope—he could not observe any such parallax and thus could not solve the problem. Instead, he drew the apparently logical but erroneous conclusion that the Earth was fixed in the universe.

Nobody at the time suspected the enormous size of the gulf between our solar system and the closest stars. It would take another 200 years before a credible parallax could be measured. Not until 1838 Friedrich Bessel announced the distance to 61 Cygni, a star in the constellation of Cygnus. His value equalled about 660,000 AU! Since many stars were much fainter than 61 Cygni, it was obvious that many of them consequently must be much more distant. It was incontestable that the distances to the stars were

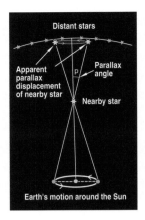

FIG. 1.2 The stellar parallax is an apparent effect caused by the motion of the Earth (Peter Linde)

enormous, and man's perception of the size of the universe was changed at a stroke.

But the revolution in understanding astronomical distances and the true scale of the universe had only just begun. Astronomers gradually acknowledged the fact that the Sun was only one out of millions of stars in a larger system of stars, called the Milky Way. Ideas along these lines had existed for a long time. The Milky Way was observable as a band of light across a dark night sky. Using his small telescope, the famous seventeenth century physicist and astronomer Galilei Galileo was able to determine that the faint light actually originated from enormous quantities of very faint stars that were not visible to the naked eye individually. In the eighteenth century the English astronomer William Herschel tried to determine the position of the Sun in the Milky Way by counting the number of stars in various directions. However, he arrived at the false conclusion that the Sun was located at the centre of the Milky Way. Today we know that this was due to the fact that dark clouds of interstellar dust obscure the majority of the stars of the Milky Way, a fact not known in the time of Herschel.

During the nineteenth century, scientists carried on a lively debate about the size of the universe. One major problem was the nature of the so-called nebulae. After the introduction of the telescope, observers realised that there was much more to the sky than stars and a few planets. Here and there, one could see diffuse patches of light of unknown nature. Some could be identified as comets, slowly moving with respect to the stars, and evidently members of our solar system. In contrast, others were fixed in position and did not move at all. Since they could be mistaken for comets, the French comet researcher Charles Messier in the eighteenth century decided to set up a catalogue of such nebulous and obscure objects. This list, which eventually contained 110 objects, became known as the Messier catalogue; the Messier objects still remain a treasure-chest for any amateur astronomer who wishes to explore the sky with a telescope.

With improved instruments, details of the objects could finally be resolved. Some objects turned out to be stellar clusters, sometimes with tens of thousands of members. Other objects, however, remained diffuse light patches, even when observed through large telescopes. These became known as "nebulae". Amongst the nebulae there seemed to be various types.

The largest contemporary telescope was constructed in 1847 by Lord Rosse. It was placed at Birr Castle, in the middle of Ireland. With a mirror of an impressive diameter of 183 cm, it could gather much more light to the human eye than had previously been possible. In spite of the less than favourable observational conditions in Ireland, Lord Rosse became the first to distinguish structures in some nebulae. He became famous for his drawings of Messier 51, which showed a clear spiral structure (Fig. 1.3).

The battle concerning the true nature of the nebulae raged throughout the entire nineteenth century and part of the early twentieth century. Many suggested that the nebulae were giant gaseous clouds that indeed were very distant, but not more so than many stars. Others proposed that the nebulae consisted of light from enormous amounts of stars, seen from a very long distance. Curiously enough, both sides were right to some extent. Meanwhile, the technological development led to new capabilities. Both photography and spectroscopy were invented in the middle of the nineteenth century, and with these new methods

FIG. 1.3 Lord Rosse's drawing of M51 from 1845. This "nebula" subsequently turned out to be stellar system, a galaxy, with several billion stars (William Parsons)

observers discovered that nebulae showed spectral lines in the same way as gases do in a laboratory. (We will take a closer look at the secrets of the spectrum in Chap. 5.) Apparently, some nebulae actually were gaseous objects while others exhibited a continuous spectrum, more similar to what you could expect from the light of many faint stars.

The Universe Expands—Again

In the beginning of the twentieth century the battle was at a decisive point—and the solution arrived. During the 1920s our perception of the universe underwent at least one more revolution. First, the size of the Milky Way and the Sun's position in it was established. The American astronomer Harlow Shapley had been studying a special type of objects for some time, the so called globular clusters. These are enormous conglomerates of stars, often containing hundreds of thousands of objects (Fig. 1.4). Approximately 75 clusters were known at this time, and using novel measurement techniques Shapley determined that they were very distant. He suggested that these star clusters were in fact situated symmetrically relative to our entire galaxy, the Milky Way. If the Earth was located at the centre of the galaxy, then the distribution of clusters in our night sky would also be equal in all directions. However, this is not

FIG. 1.4 The globular star cluster M80. Hundreds of thousands of stars are orbiting a common centre (Hubble Heritage Team (AURA/STSci/NASA))

the case; the clusters are unevenly distributed in our sky with most of them visible in a certain area. This meant that the Sun—and hence the Earth—were consigned to a much more obscure position closer to the periphery of the Galaxy. At the same time it became evident that the Milky Way itself must be much larger than previously believed.

Somewhat later another sensational discovery was made. Edwin Hubble used the recently inaugurated giant telescope at Mount Wilson in California to study the Andromeda nebula. This telescope, with a mirror diameter of two and a half metres, was now the largest in the world and completely superior to the telescope of Lord Rosse. Hubble's research also benefited further by spectacular technology advancements for astronomical observations. Hubble was able to resolve the Andromeda nebula into individual stars and to determine the distance to them. The Swedish astronomer Knut Lundmark simultaneously made similar observations

FIG. 1.5 The Andromeda galaxy—a stellar system with more than 400 billion stars (Adam Evans)

and arrived at the same conclusion. The Andromeda nebula was, in fact, a galaxy, a gigantic stellar system in its own right and certainly not a gaseous nebula (Fig. 1.5). It was clear that the Milky Way as such was not the entire universe, but instead was a single galaxy among thousands of others. The universe was incredibly larger than anyone had hitherto imagined.

A Suitable Yardstick?

Most people today know that the universe is large. But nobody can in earnest comprehend its actual size. Common objects like a car or a house are well-known concepts which are naturally within our intellectual grasp. But when the scale increases, the comprehensibility decreases. A common way to try to understand the incomprehensible is to use analogies or to make models, often by using some kind of yardstick. One such yardstick to illuminate the astronomical scale concerns light and its velocity.

The nature of light and whether its speed of propagation was finite or infinite was, for a long time, one of the major scientific mysteries. But as early as 1676 the Danish astronomer Ole Römer succeeded in making a measurement of the speed of light. By this time both Kepler and Newton had considerably increased the understanding of how the celestial bodies moved, and Newton's law of gravitation could be applied everywhere in the solar system. Among other things, it was clear that the satellites of Jupiter were moving in a regular and predictable way. Römer made accurate observations of these motions during a longer period of time, noting especially how they moved in and out of Jupiter's own shadow. After some time he noted a gradual shift between his observed timings and the predicted timings of satellite eclipses. He arrived at the correct conclusion that the difference was due to the fact that the distance between the Earth and Jupiter had changed between his observations. This meant that the speed of light must be finite, and by using the available estimation of the distance between the Earth and the Sun he made the first calculation of the speed of light.

The modern value of the speed of light is nowadays determined to 299,792.458 km/s. This is by definition exact, since the definition of a metre since 1983 is directly linked to the speed of light.

In the astronomical context it rapidly becomes cumbersome to measure distances in everyday terms like metres or kilometres. Millions of kilometres can be used, marginally, within our solar system, but beyond that even the closest stars are so far away that the numbers become truly "astronomical". We need a larger yardstick and instead we use "light time", i.e. the distance that the light traverses in a given time unit. The concepts of light second, light hours and light years give us new units, making it possible to more readily understand the distances, or at least their proportions. One light second thus becomes the distance light travels in one second, which corresponds to almost 300,000 km. A light minute is 60 times longer than a light second and so on. Consequently, a light year corresponds to 31,557,600 light seconds.

With this new yardstick, let us try to illustrate the distances. The distance to the Moon (Fig 1.6) is on the average 384,000 km and thus corresponds to a little more than a light second. The speed of light became a tangible effect when Neil Armstrong stepped down on the lunar surface for the first time, since the

Fig. 1.6 The Moon lies at a distance of one light second from the Earth (NASA/Sean Smith)

radio communication with the Houston space centre was always delayed by at least two seconds—the time between a question and an answer.

However, the distance between the Earth and the Sun is considerably longer, corresponding to eight light minutes. Within our solar system the distances vary on a scale between light minutes and light hours. The eight main planets lie within a diameter of 10 light hours (Fig. 1.7). Beyond them there are other remains of various kinds from the creation of the solar system, the total extent of the solar system can be estimated to about 100 light hours.

It is worth noting that the research space probes that have been sent out into the solar system, among others to Jupiter and Saturn, have needed several years to complete such voyages. It also follows

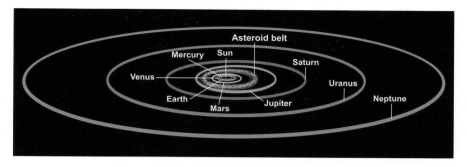

Fɪɢ. **1.7** The orbit of the outermost planet, Neptune, has a diameter of nearly 10 light hours (Peter Linde)

that distant probes, in large part, need to be able to act on their own, since a command from Earth may take hours to reach it.

As has already been indicated, the step out to the stars is indeed huge. Light hours are not nearly sufficient for such distances. To the nearest star, α Centauri, we count 4.3 light years. Keep in mind that there are almost 9000 hours in a year. Obviously, light years correspond to an enormous distance. Measured in kilometres it would be 9½ trillion, a number which of course is graspable neither by ordinary persons nor by astronomers. In a later chapter the dream of traveling to the stars will be examined in some detail. Suffice to mention now that a space craft made by humans right now is currently on its way out into the interstellar void and that it so far has completed 0.002 light years...

Since Bessel's ground-breaking determination of a stellar distance, advancements have continued rapidly. About 150 years later, the spacecraft Hipparcos achieved fundamental further advances when it made very advanced positional measurements of about a million stars in the solar neighbourhood. Among other things, precise distances were measured for all stars within 400 light years. The Hipparcos success is currently being followed up by an even more advanced astrometric satellite, the Gaia. About a billion stars will be measured with a precision exceeding that of Hipparcos at least 100 times. More on this in Chap. 7.

Today we have a rather well-founded view about stellar distances. Obviously, the stars we see by naked eye are relatively close, measured using this yard-stick (Fig. 1.8). The remotest naked eye stars, which are in reality intrinsically very bright,

FIG. 1.8 Most stars visible by naked eye are between 4 and 1000 light years away. The stars in the Plough (aka Big Dipper) are about 100 light years away (Peter Linde)

lie at a distance of about 1000 light years. Such stars can output more than a thousand times the energy of our Sun and are, therefore, easier to detect at longer distances.

Even using very large telescopes, there are limits to what we can see in our own galaxy, the Milky Way. As previously mentioned, between the stars space is not completely empty. In places, thin matter drifts in the form of gas and dust which obscures the line of sight. This problem becomes worse with increasing distance, particularly when looking in the direction of the plane of the Milky Way. That is why ordinary light cannot reach us from distant parts of the Galaxy. But radiation in longer wavelengths, like infrared or radio radiation, penetrates more easily through the dust. Observations in these wavelengths have enabled researchers to create a successively more detailed picture of the Milky Way (Fig. 1.9). The diameter of its visible matter has nowadays been determined to be about 100,000 light years. The Sun belongs to one of its spiral arms at a distance of about 25,000 light years from the centre.

Sun's position

FIG. 1.9 This illustration shows the modern conception of our own galaxy, the Milky Way, seen from above. The diameter is about a 100,000 light years. It contains more than 200 billion stars as well as large quantities of gas and dust. The location of our solar system is marked (NASA/JPL Caltech)

Unfortunately, the concept of light years also becomes unwieldy the further out in the universe we are looking. Additionally, we have a problem with the very definition of a distance. Do we mean the distance of the object when its light reached us? Or the distance it is at the moment we are looking at it? Even if we use light years as a measure of distance, the time aspect can hardly be neglected. We get problems with the idea of simultaneity. If we observe a star at a distance of 1000 light years, the light has needed 1000 years to reach us. Meanwhile, the star may very well have moved away—and has in addition become 1000 years older. This is not a major problem when we look at most objects inside our own galaxy. But when we look even further out into the universe, this effect becomes much more important. In fact, it then probably becomes more relevant to express oneself in terms of time than in distance, because then we are talking about millions of light years—and of years.

The Universe on Its Largest Scale

Hubble succeeded in determining a preliminary distance to the Andromeda galaxy. He continued with another revolutionary discovery. By studying spectra from more distant galaxies he could draw the conclusion that all distant galaxies are receding away from us. The further away, the faster they recede. This was interpreted as evidence of an expansion of the whole universe, and is now considered a cornerstone in our present understanding of how the universe was created, the theory of the Big Bang. At the same time, our perception of the real size of the universe became somewhat clearer.

The modern estimation of the distance to the Andromeda galaxy is about 2.5 million light years. With the exception of a few minor satellite galaxies of the Milky Way (most notably the Magellanic Clouds), this is our closest galaxy. With good visibility it can be seen by naked eye and is normally considered the most remote object visible without aid. So, the light we see now was transmitted 2.5 million years ago—but the perspectives rapidly become even more staggering. We now believe that the structure of the universe is hierarchical, a little bit like a Russian doll in reverse. The stars form clusters which in turn are parts of large congregates of stars, known as galaxies. Groups of galaxies form clusters. Our own galaxy and the Andromeda galaxy, together with a few minor galaxies, form the small Local Group. It, in turn, is now considered to be part of the giant Virgo cluster which contains thousands of galaxies, at a distance of about 60 million light years.

The galaxy clusters, in turn, are normally parts of so called super galaxy clusters. Figure 1.10 is based on recent observations and shows the mass distribution of the universe in the form of super clusters within the closest 2.5 billion light years. An estimation of the number of galaxies inside the observable universe is at

FIG. 1.10 The distribution of the matter of the universe within the nearest 2.5 billion light years. Every *dot* corresponds to a galaxy cluster. Small surfaces correspond to super galaxy clusters. The *colours* represent distance, *blue* is closest, *red* most distant. In the foreground our own Milky Way is seen (NASA/IPAC)

80 billion. From the figure it is apparent that the mass distribution is not even, but is interspersed by large voids.

At this level, it evidently becomes difficult to separate time and distance. A fascinating illustration is shown in Fig. 1.11. This is the deepest image that has ever been taken and represents a cross section of the universe in time and space. Only galaxies can be seen, everything from relatively well defined ones down to small diffuse dots of lights. The closest galaxy is at a distance of about 1 billion light years away while the furthest are at the very limit of the observable universe. Considering that the light from such sources was transmitted more than 10 billion years ago, one realises, after some thought, that, since the universe is continually expanding, these galaxies are much more remote "today" and that they surely look different "now".

A small calculus example may illuminate this: the age of the universe according to current theories has been determined to 13.8 billion years. That long ago the Big Bing took place and our universe was created. Since then the universe has continually expanded. With the largest available telescopes it has so far been possible to observe objects that existed as early as 0.5 billion years

FIG. **1.11** An image cut from the Ultra Deep Field of the Hubble telescope. The size of the field corresponds to the angle subtended by a golf ball seen at a distance of 200 m. All visible objects are galaxies. The smallest are located close the limit of the observable universe (NASA)

after the Big Bang. The distance then evidently was 13.3 billion light years. But since then the universe has expanded over the course of 13.3 billion years. This means that these objects from the early time of the universe are even more remote today, estimated at a distance of 31.5 billion light years. In Box 1.1 we summarise our yardstick based on the concept of light time.

Mankind at Its Place in the Universe?

Our understanding of the size and characteristics of the universe has drastically influenced mankind's view of its own role and place in the universe. Until the Middle Ages it appeared obvious that the Earth was the centre of the universe. Man was the lord of the creation, or at least its caretaker. The possibility of intelligent life elsewhere under such circumstances was not very probable and certainly not very attractive. As late as the seventeenth century, the Catholic church still banned advocates of a more modern way of thinking. Giordano Bruno became one of the more famous victims. He claimed that the universe was infinite and that it contained infinitely many worlds and that all these were populated by intelligent beings. He was also at odds with other Catholic views and, as a consequence, was burned at the stake in the year 1600.

In 1633, Galileo Galilei had to denounce his heretic theories about the Earth not being at the centre of the universe. Not until the year 2000 was this sentence withdrawn by the Catholic church. Quite a lot had by then happened to our perception of the universe.

Maybe now, humanity—for the first time—is ready to seriously accept the possibility that alien civilisations may exist, many perhaps more advanced than our own. At the same time we have the insight that distances in space are huge and the difficulties to communicate through space and time are enormous.

Box 1.1: Measurement Units in Space

1 light second = from here to the Moon
100 light hours = diameter of our solar system
10 light years = our closest stars neighbours
100,000 light years = diameter of our galaxy, the Milky Way
10 million light years = our closest galaxy neighbours
10 billion light years = towards limits of observable universe

2. How Is a Planetary System Formed?

An important prerequisite for the existence of life is that planets exist on which life can develop and thrive. Until recently, our own solar system was the only one we knew about. It is still, by far, the one that we know best. How did it form?

During the seventeenth century the modern view of the structure of our solar system finally broke through. Observations of Tycho Brahe and the subsequent analysis of Kepler enabled Newton to formulate his theory of gravitation, and a modern foundation for the understanding of the universe was born. New discoveries expanded our perception of our own solar system. William Herschel, famous German-British astronomer, discovered the planet Uranus in 1781. The theory of gravitation triumphed with the discovery of the planet Neptune in 1846, as a direct result of advanced mathematical calculations, inspired by the deviations found in the movements of Uranus.

Throughout the centuries a variety of theories trying to explain the creation of our solar system have been discussed. But when a more modern view of the universe was eventually established, theories became more realistic. The basic assumption was that the solar system, with its sun and planets, had been formed by way of a flat, rotating disk of matter. This accounts for the fact that planets lie in the same plane and orbit in the same direction around the Sun. Among the first to propose such a nebula hypothesis, in the beginning of the eighteenth century, was the Swedish scientist and philosopher Emmanuel Swedenborg. The theory was further developed at the end of the same century by the German philosopher Immanuel Kant and the French astronomer Pierre-Simon Laplace.

© Springer International Publishing Switzerland 2016
P. Linde, *The Hunt for Alien Life*, Astronomers' Universe,
DOI 10.1007/978-3-319-24118-0_2

Our Solar System of Today

The nebula hypothesis agrees well with the current perception of how a planetary system is created, but in modern times further research has extended and complicated the picture. We nowadays know that there are more components in the solar system. A large number of smaller bodies has been detected and are designated as a new class of objects, the dwarf planets. Most of these newly discovered objects are located at great distances from the Sun. At least two belts of even smaller objects, minor planets or asteroids, exist in the solar system. A summary characterisation of the modern picture of our own solar system is illustrated in Fig. 2.1, and contains the following items:

- The inner planets are small and compact, mainly consisting of rocky minerals and heavy metals, like nickel and iron. Examples are Mercury, Venus, the Earth and Mars.
- The outer planets are giant gas planets, where the dominating components are hydrogen and helium, much like the composition of a star. Typical examples are Jupiter and Saturn. The smaller giants Uranus and Neptune, however, also have a larger fraction of ice and are sometimes characterised as ice giants.
- As mentioned, the planets lie roughly in the same orbital plane and move in almost circular orbits in the same direction around the Sun.
- There are remnants of various kinds left from the creation of the solar system, both smaller and larger objects. The asteroid belt contains many thousands of minor plants between Mars and Jupiter. Outside Neptune we have the Kuiper belt, a similar zone of smaller objects.
- A number of bodies, normally with diameters between 1000 and 2000 km, have been discovered and have prompted the definition of a new category of objects, namely the dwarf planets. Observers expect to find many more such distant and faint objects far out in the solar system. Both asteroids and dwarf planets can be considered to have been planetary building material that was never used.
- Still further out, about a thousand times the distance to the Kuiper belt, a spherical envelope consisting of millions of smaller bodies of ice is assumed to surround the solar system.

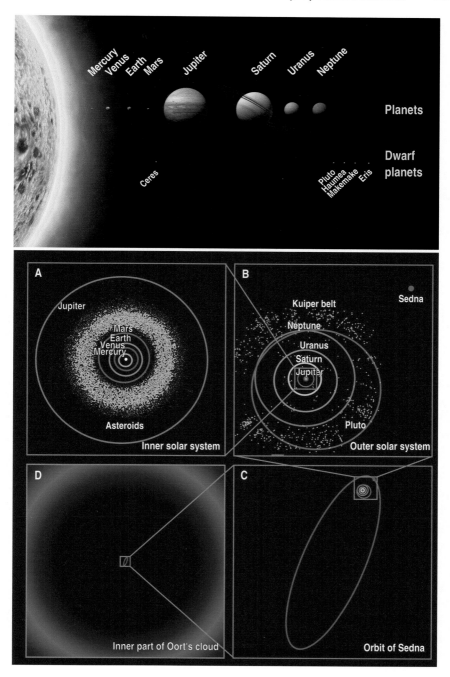

FIG. 2.1 Overview of the solar system. Top panel show planet and Sun sizes to scale, panels A-D show the respective distances to scale. The scale is successively changed, with A showing the inner parts and D indicating the Oort's cloud, assumed to be the source for comets (*Upper panel* NASA, *lower panel* Peter Linde/NASA/JPL-Caltech/R. Hurt)

This is known as the Oort Cloud, and is most probably the origin for many of the comets that occasionally approach the Sun from various angles and with strongly elliptical orbits.

- 99.86 % of the total mass of the solar system is contained in the Sun itself. Of the rest, about 90 % is contained in the masses of the giant planets Jupiter and Saturn.

These are the basic facts that have to be explained by a realistic theory for the creation of the solar system. Size and distance ratios are illustrated in Fig. 2.1. As example of a newly discovered object, the orbit of Sedna is shown. This is a small planet—probably to be classified as a dwarf planet—with a very wide orbit. Sedna needs about 11,000 years to make a single orbit around the Sun. As a comparison, the outermost main planet—Neptune—just needs 165 years for one revolution.

The Recipe for a Planetary System

To form a planetary system you primarily need raw material that also contains heavier basic elements that can form dust of various kinds. However, such elements did not exist in the universe from the very beginning. The original matter that was created shortly after the birth of the universe essentially only contained hydrogen and helium, the two simplest basic elements. The elements that dominate the Earth, like carbon, oxygen and iron, were formed later. These elements result from stars through nuclear processes. At very high temperatures in their cores, stars burn their hydrogen—and eventually their helium. A similar process takes place when a hydrogen bomb explodes. By means of fusion, simple atom nuclei are smashed together into more complicated nuclei. As an example, three helium nuclei may form a carbon nucleus. Astronomers simplify the terminology by designating all elements heavier than helium as metals. Through various fusion processes, basic elements up to iron can be formed inside normal stars. Heavier elements than that do not yield any energy; instead, energy has to be supplied in order for such elements to be formed.

The first generation of stars only contained hydrogen and helium. It is believed that these stars had very large masses and, after a short life, generally ended explosively and violently. In this way, the newly formed heavier elements inside the star could escape out into interstellar space as nebulous debris. Much later this material, in turn, could form new stars which were now enriched with heavier elements. This process has been repeated in successive stellar generations through billions of years, with the result that stars of later generations have a much higher fraction of heavier elements in their original composition.

No stars from the very first generation have so far been observed. They do not exist in our neighbourhood (or time) but their light, transmitted when the universe was still young, may still be on its way towards us from very distant parts of the universe. Perhaps observations of these stars may become possible with the new giant telescopes now under development. What we can see today are stars from later generations. Our own Sun is a comparatively metal-rich star of a late generation.

The theory of how a star and its planetary system are formed is now reasonably well developed, but many details still remain to explain. Initially, a large group of stars (a cluster) forms out of giant gas clouds. We observe these as nebulae (a beautiful example is seen in Fig. 2.2), generally several light years in size. Initially small and random motions in the cloud—gas turbulence—may at some point be affected by shock waves from some nearby unstable star or even by a supernova explosion in its neighbourhood. When this happens, the matter is locally compressed. Through the influence of increased gravity, these points can continue to collect matter which starts a rapidly accelerating process of accumulation. As a result, density increases in such places, which in turn increases temperature and pressure—which creates an outwardly and thus balancing force. Concurrent movements lead to the effect that the gas and the dust are concentrated into a rotating disk, a so-called accretion disk. Such disks have now been directly observed in some gas nebulae. In Fig. 2.3 we see such disks in the Orion Nebula as discovered by the Hubble telescope. The future normal star—the proto-star—initially will be rotating very fast, and in order to maintain its stability it needs to decelerate in some way. This probably happens through gas friction (viscosity) against its closest surroundings. When the pressure and temperature at the centre have risen to a sufficient level—several million degrees after about 100,000 years—nuclear fusion reactions start, becoming the continuous power source for the star.

FIG. 2.2 Nebula in the constellation of the Eagle. The gas and the dust are the basic building materials for the creation of new planetary systems. In the *upper part* is seen the light from recently formed stars (NASA)

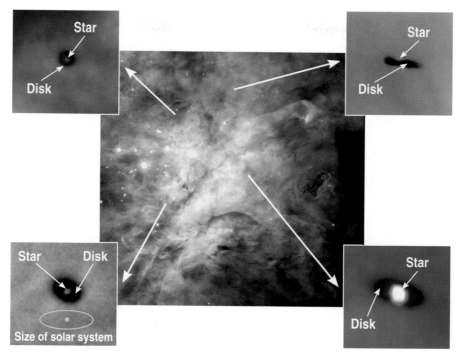

FIG. 2.3 The Hubble Space Telescope could in 1995 image disks inside the Orion nebula. From such disks new planetary systems can form (NASA/ Peter Linde)

The Final Stages of Planetary Formation

In the remaining parts of the proto-star disk planets can form. In addition to hydrogen and helium, the disk contains various kinds of dust and ice particles. Figure 2.4 shows an artist's conception of how such a disk may look, and Fig. 2.5 shows a possible view of a planetary system being formed. The disk originally has a high temperature which affects this material in different ways. In the inner parts most existing chemical compounds are vaporised but in the outer parts, water and other volatiles remain in frozen form as ice particles. The limit for this zone is called the snow line. Inside the snow line water can only exist as vapour, and this vapour can be affected by a particle wind from the newly formed and active star. Water vapour cannot form into planets. The disk probably remains for about 3 million years. The dust in the disk undergoes a process

Fig. 2.4 An artists conception of how a planetary system can be created. A rotating disk of gas and dust is successively condensing. In the middle a proto-star is born (NASA)

Fig. 2.5 Between 10 and 100 million years later, a planetary system is in formation. Smaller and larger stones and rocks collide and successively grow into larger bodies (Fahad Sulehria/www.novacelestia.com)

where the originally very small grains (about 0.001 mm) start to conglomerate into larger sized grains. These continue to grow to a size of a few centimetres. Using advanced computer simulations, astronomers are studying what happens next. It turns out that the small pebbles spontaneously can form planetesimals if they are subjected to a fluid medium, in this case the surrounding gas. The size of these planetesimals can reach a thousand kilometres. The phenomenon is due to the fact that this growth process actually conserves energy, in the same way that cyclists save energy by moving closely together in a group, or when birds fly in a V-shaped formation.

The inner parts of the stellar disk contains mainly minerals and metals that can form planets. The planetesimals form through violent collisions, and their gradually increased gravitation attracts larger and larger bodies. Again via advanced simulations, astronomers have been able to track down many of the details of this process. After about 100 million years this inner part of the disk is cleared and a small number of rocky planets have formed. However, recently a remarkable image obtained with the new microwave telescope ALMA (Fig. 2.6) suggests that this formative time may, in some cases, be considerably shorter.

FIG. 2.6 This image from ALMA, the Atacama Large Millimeter/submillimeter Array, reveals extraordinarily fine detail in the planet-forming disc around a young star (ESO/NAOJ/NRAO)

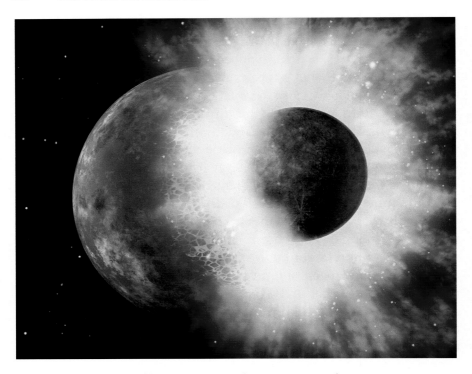

FIG. 2.7 About 4.5 billion years ago the current Earth-Moon system was probably formed as a result of a gigantic collision between two large proto-planets (NASA)

Evidently, this early period was characterised by violent collisions between many larger bodies. About 4.5 billion years ago the Earth-Moon system was apparently formed when a large (perhaps about 5000 km sized) body collided with the proto-earth (Fig. 2.7). About 3.8 billion years ago the proto-planets in our inner solar system seem to have been the subject of an intensive bombardment. The period is known as the "Late Heavy Bombardment". It is at this time that the Moon obtained most of its craters. Evidence for this event was brought back by the Apollo astronauts from their samples of the lunar surface. Simultaneously, the other inner planets got their share, including the Earth. While traces from this period are mostly obliterated on Earth and Venus, many scars remain on Mercury and Mars. It is not yet established why the bombardment was so intense at this specific time. Perhaps some of the outer giant planets at the time for some reason modified their orbits and gravitationally affected the asteroid belt, redirecting many asteroids on collision course with the inner planets.

The creation theory of the giant planets is currently less developed. Their cores are formed from planetesimals in a way similar to the inner planets. Since they are located outside the snow line, many planetesimals consist of ice instead of rock. If they accumulate about ten earth masses of ice they can then continue to attract large amounts of the surrounding disk hydrogen and helium, and become gas giants with perhaps a mass 300 times larger than the Earth's. In our solar system, Jupiter and Saturn are examples of this. Giants with smaller cores attracts less amounts of gas and become what essentially are ice giants. Uranus and Neptune are of this type, with about 15 times the mass of the Earth.

The details about where the giant planets are formed, how they possibly move around and how they affect their surrounding through gravitation are important questions that are related to the possible emergence of life on a planet. Through studies of planetary systems of other stars it should be possible to improve our understanding of these processes. We will return to this further on.

The Heaviest Elements

It is difficult to leave this topic without expanding a little on the remarkable way the heavier elements are formed. As mentioned, basic elements heavier than iron need an external energy source in order to be created, which means that they cannot be created under normal circumstances in ordinary stars. Yet, such elements are common on Earth. Some examples are tin, lead, silver and gold. In fact, many of these elements are constituents in what today is carried as jewellery. How have these elements ended up here? We know that nuclear reactions necessary to form such elements require extremely high temperatures. These can only be obtained in incredibly violent processes in the universe. One example of this is the supernova explosion. Some stars, mainly those with an original mass larger than the Sun's, become unstable towards the end of their lives and annihilate themselves in enormous explo-

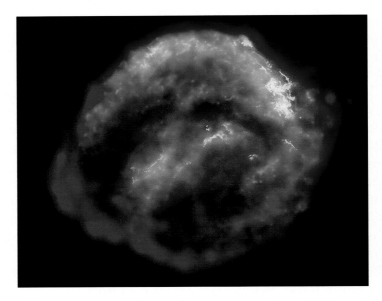

Fɪɢ. **2.8** The remnants of the Kepler supernova from 1604 as they look today. The image is a composite from observations made in optical, infra-red and X-ray radiation. The latter are illustrated using colours (NASA)

sions. Their brightness becomes a billion times increased and the temperature at the core can rise to 100 billion degrees! However, supernovae are rare. The latest witnessed supernova in the Milky Way was observed in 1604 and is known as Kepler's supernova. The remnants from this are seen in Fig. 2.8, which gives a clear impression of an enormous explosion.

In spite of its rarity it is clear that the elements that we find on the Earth, and to some extent, inside our own bodies, once must have been created in supernovae, probably distant in both space and time. That we ourselves are literally made of star stuff is a fascinating perspective which shows the close connection between life on Earth and the events out in the universe.

3. The Development of Life on Earth

Many dream that life is not something unique to our Earth, that mankind is not sentenced to loneliness in the universe, that there are, in an unknown place somewhere, thinking and feeling beings with whom we can communicate and share knowledge and experiences. But so far, our own planet is the only place where we know for sure that life exists. Life on Earth expresses itself in many forms and can be found in the most unexpected places. It seems there is an almost overwhelming force that drives life to develop in many directions at the same time, filling all the available niches. The number of different species is enormous. A modern estimation shows that the Earth contains close to ten million species, in addition to many more primitive life forms. Of these, only two million are known, although thousands of new species are discovered every year.

Does this mean that life begins wherever there is the slightest opportunity, even under circumstances that are very different from what we now have on Earth? That life is common also out in the universe, perhaps on millions of planets? We don't know—yet. Let us begin by analysing what makes life such a success here on Earth. Earth life will provide important clues about how and where to search for alien life.

Earth: A Special Planet?

The geology of the Earth has a few specific properties. Although these are not unique in the solar system, on our planet they do coincide in a way that may have been decisive for the emergence and development of life. One of these qualities is the obvious fact that the Earth has large quantities of liquid water. Less apparent, but still very important, is the fact that the Earth has volcanism, plate tectonics and a magnetic field. To understand how these properties work and interact we need to take a look at the composition and structure inside the Earth.

© Springer International Publishing Switzerland 2016
P. Linde, *The Hunt for Alien Life*, Astronomers' Universe,
DOI 10.1007/978-3-319-24118-0_3

The age of the Earth is estimated at about 4.5 billion years. 4.5 billion years ago, the Earth's formation process was coming to an end, as we have seen from Chap. 2, in violent collisions and at high temperatures. When the raw material of the Earth accumulated in the formation process, a stratification took place: heavier elements sank deeper down into the Earth's core. An indication of this is the fact that the density of the entire Earth is about 5.5 g/cm^3, while the average density of its outer regions is only about 3 g/cm^3. Evidently, the inner part of the Earth consists of heavier elements. Our knowledge about the inside of the Earth is not very detailed, but some observations have been possible to do, for instance during earthquakes. These give an opportunity to study how the tremors and shock waves are propagated through the Earth in different ways, which in turn gives information about the inner composition and density of the planet. Researchers have concluded that the core consists of a mixture of iron and nickel (Fig. 3.1). It has a

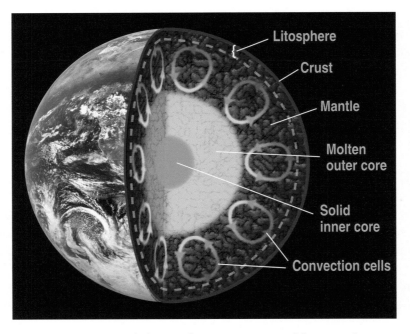

Fig. 3.1 The inner part of the Earth contains several layers. The temperature increases inwards and the texture of the matter varies. The movements in the mantle explain the slow displacements of the continents (Peter Linde)

solid inner part with an estimated temperature of about 5500 °C and a more liquid surrounding part with a temperature about 1000° lower than at the centre. That the centre is still solid, in spite of its higher temperature, is due to the enormous pressures that prevail there. Outside the core we have the mantle, which can be likened to a liquid but sluggish mass of various forms of minerals and rock materials. Due to the difference in temperature between the upper and lower parts of the mantle, a circular motion in the mantle is created in the form of convection cells. Convection is a form of heat transfer through currents in a medium that can either be a fluid or a gas. A temperature gradient between the lower hotter level and the cooler upper level causes this motion. The lower level expands due to the temperature, thus creating a lower density. This gives the matter buoyancy with regard to its immediate surroundings. The cooling that takes place at the upper level increases the density again and the liquid or gas falls back. Among many examples of this phenomenon in our daily life, we can simply observe a boiling liquid in a kitchen pan. In a slowly moving liquid, like oil, it is easy to see how up-and-down motions are created. Another example of slow convection can be seen in the ever-popular lava lamps.

Furthest away from the Earth core lies the crust. It is a remarkably thin layer mainly consisting of various rock composites, originally of volcanic nature, which, through various geological processes, were modified into other types of rock. Under the oceans the thickness is only 5–10 km, while under the continents it is somewhat thicker, 30–50 km.

Plate Tectonics: The Continents Move

Since the stabilisation of the Earth, about 4 billion years ago, plate tectonics has played a major role. The crust of the Earth is broken into a number of giant plates which are affected from below by the convection motions within the mantle. This makes them move, much like material on a conveyor belt. A visible sign of this is the continental drift, which shifts the continents relative to each other, normally a few centimetres per year. In the border zones, where the plates meet or separate, there are usually dramatic geological upheavals, although at a slow rate. Mountain ranges are created in this way. A prime example is the Himalayas, which is still growing since the Indian continental plate is colliding with the Euro-Asian plate. There are also places where the plates are separating, for example in the middle of the Atlantic floor where new crust is developed from beneath, filling in the increasingly growing separation. The other end of a sea-level plate usually collides with a continental plate in such a way that it is forced down under the continental plate, back into the mantle again. During this process, the interfacing part is heated, leading to increased volcanic activity. The movements of the continental plates are of major importance for the conditions for life on Earth, as we shall soon see.

A Magnetic Field Protects against Dangerous Radiation

Anyone who has used a compass knows that the Earth has a magnetic field. But how is the magnetic field of a planet actually created? Three conditions seem to be decisive. Firstly, you need an electrically conductive and floating medium in the inner part of the planet. Secondly, this medium must have convective motions, and thirdly, the whole planet must rotate at a certain speed. Electrical currents are created much like those in a dynamo. These, in turn, generate the magnetic field. The Earth clearly fulfils these conditions, and as a result has a rather strong magnetic field. As we shall see in the following chapter, this characteristic separates it from its closest neighbours.

The magnetic field constitutes an important protection for the Earth and for the organisms on it. The field has the property of deflecting electrically charged particles coming from space. Without this protection, eruptions from the Sun and from powerful cosmic radiation would lead to a gradual thinning of the atmosphere. Additionally, the radiation level at the ground would probably be lethal for most forms of land-based life.

The Earth Is a Water Planet

The fact that the Earth has large quantities of liquid water may be considered as the most important requirement for the emergence and development of life. Water has a number of important chemical properties. Firstly, it is an excellent solvent- and transportation medium, which means that minerals are easily accessible for necessary chemical reactions at a molecular level. Secondly, compared to many other liquids, water remains liquid over a rather large temperature interval, as is well known, from 0 to 100 °C (at normal atmospheric pressure). Thirdly, water has a high heat capacity—it takes a lot of energy to increase its temperature, which in turn may be kept for a long time. This means that water in large quantities can have a strongly climate stabilising role. Water also has the unusual property of decreasing its density when frozen. Ice floats on water, which means that lakes or oceans will not easily freeze solid. Aquatic life therefore can survive under ice during long periods of cold. The water also protects living organisms from harmful radiation, something that was especially important during the early period of the Earth.

The origin of the water on Earth is not entirely clear. There are two main theories, and it is possible that both in combination explain the large quantities of water on Earth. One theory says that when the Earth was formed, large amounts of lighter elements were trapped in the crust, among them various gases and water. During the earliest epochs much of this material was released through violent volcanic activity, forming the basis for the first oceans and the first atmosphere. The second theory says that water was added to the Earth somewhat later through collisions with comets and protoplanets rich in ice. If this theory is closest to the truth, it could mean that the existence of water on the Earth is a coincidence. The isotopic content of the water, i.e. which fraction of deuterium to hydrogen it contains, can give clues to its origin. Data for our ocean water is not in agreement with data from several comet measurements. For instance, recent results from the Rosetta landing on comet 67P/Churyumov-Gerasimenko gave a much higher ratio than for terrestrial ocean water. This strongly indicates that the water on Earth cannot have come from some comet families, but is instead believed to have its origin in icy asteroids. In any case, it seems probable that the oceans were formed at a very early stage in the existence of the Earth.

The Green House Effect: The Basis for Life on Earth

In today's social debate, the increasing green house effect is a source of major concern. There is evidence that the average temperature of the Earth increases with changes in the global climate as consequence. Human activity in the form of release of large quantities of carbon dioxide threatens to accelerate the process. It may then sound a bit paradoxical to point out that the green house effect in reality is the basis for all life in Earth. If you calculate the average temperature Earth would have without an atmosphere, thus depending only on the distance to the Sun, the answer would be about +5 °C. But the Earth has an atmosphere, and it reflects back about 30 % of the solar energy. Hence we would instead end up considerably below the freezing point of water, at about 15–20° below. But, as a matter of fact, the average temperature is about +15 °C. This crucial difference is explained by a natural and ongoing green house effect. The green house effect thus facilitates benign conditions for life, but that is not all—it also has a regulating and stabilising effect (see Box 3.1).

Box 3.1: How Does the Greenhouse Effect Work?

The greenhouse effect begins by sunlight reaching the surface of Earth and heating the ground. Some heat is trapped, because while the atmosphere can transmit visual light, it is opaque for the infrared heat radiation that the ground re-radiates. When an atmosphere contains gases like water vapour, methane and carbon dioxide, this blocking force increases. The greenhouse effect gradually increases the temperature until a higher level is reached at which a new equilibrium settles between the total input and output radiation (Fig. 3.2).

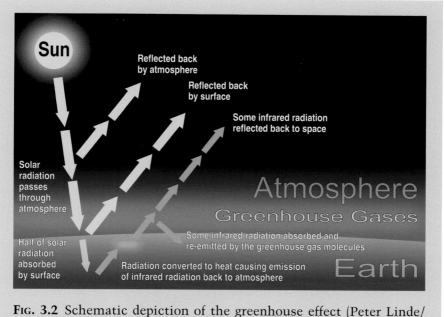

Fig. 3.2 Schematic depiction of the greenhouse effect (Peter Linde/ ZooFari)

The Earth Has a Built-in Thermostat

Even if today's worries about the effects of global warming must be taken very seriously, one should be aware that the climate of the Earth has changed drastically many times before. Since the Earth was created, the radiation of the Sun has increased somewhat, the orbit of the Earth and the inclination of the Earth axis has changed a little, and so on. In spite of such influences, the climate on Earth somehow always seems to recover in the long run, both from unusually warm and from unusually cold periods. We find an important explanation in the cycle that governs the level of carbon dioxide in the atmosphere, which operates like a thermostat. In Fig. 3.3 we see how it works. The amount of carbon dioxide which exists in the atmosphere can be reduced by

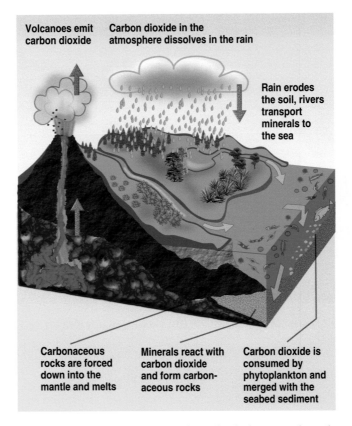

Fig. 3.3 The circulation of carbon dioxide balances the climate (Eva Dagnegård)

dissolving in rain drops that fall to the ground. Rain also leads to erosion where rivers carry dissolved minerals towards the seas. The minerals react with the carbon dioxide and a sea bottom is freshly created from carbonaceous minerals. Due to the movements from plate tectonics, this newly created sea bottom eventually is pushed down below a continental plate. At that point the material is heated up and can be returned to the atmosphere via volcanic activity. The regulating effect of the cycle arises from a phenomenon called negative coupling. At high carbon dioxide levels in the atmosphere, the temperature rises due to the increased greenhouse effect. This leads to increased precipitation, which then decreases the amount of carbon dioxide, after which the temperature decreases again.

Should the Earth be subjected to a colder climate by external forces, the mechanism works the other way around. Then the carbon dioxide level increases which in turn increase the greenhouse effect. This mechanism is not the only one which creates a regulation of the climate, but it is the most effective one. However, the problem is that it needs very long time, millions of years, to make any difference. So evidently it does not present a solution to more short-term and immediate climate problems.

How to Recognise Life?

Before we delve into the questions of how life was created, we must start by trying to establish a definition of life. A lot depends on what you include in the concept of life. It soon becomes a tricky question with philosophical and perhaps religious implications, which do not have straightforward answers. Not even the definition of highly developed "life" is simple. We certainly associate phenomena such as intelligence and awareness with life. But even these two characteristics are difficult to get a grip on. Is the human being something more than just the sum of all its atoms? From a purely functional point of view, man has already surpassed all its physical properties through technological advances in different ways. With the enormous advancement of computer science, the prospect of computers surpassing man with respect to intelligence and awareness approaches rapidly. Looking at the other end of the scale, at primitive life forms, it is again not easy to draw the line for what life is, but for other reasons. Nevertheless, the following short list is an effort to define the concept of life.

- Life is characterised by a high degree of order and organisation
- Life reproduces in some way
- Life grows and develops and thus needs an energy source
- Life is controlled by heredity
- Life reacts to its environment and adapts to it

The last condition we recognise as part of the Darwinian theory of evolution. When Darwin published the theory in 1859, it was certainly controversial in religious circles (and in extreme cases still seems to be) but was immediately accepted by a majority of contemporary researchers. But neither Darwin nor anybody else could at the time give a fundamental explanation of how species were created and developed. Another century would pass before microbiology in a revolutionary way revealed the relevant underlying mechanisms. The requirement for order and organisation was overwhelmingly fulfilled in the fantastic world that was exposed with the help of microscopes and other research tools. The DNA code hidden inside every living cell explained reproduction and heredity. It was realised that life could both develop and adapt through modifications of the DNA code, by means of mutations in the organism. The pressure from the environment, together with modifications of the genetic code, were the keys to how new species could be created, as old species had to adapt or become extinct.

The Miracle of Life

To go into in the fascinating details of how life works at a cellular level is unfortunately outside the scope of this book. We need to be satisfied by stating that the organisation and complexity at the microlevel is simply fantastic. It is hardly surprising that some people see a "designer" in all the delicate mechanisms that are needed in order for life to function. The rest of us remain amazed by the incredible power of evolution which, during thousands of millennia, has created so many miraculous functions and organisms.

We will, however, comment on some fundamental properties of life. Earthly life is carbon based—but what does this actually mean? If we take a closer look at the basic elements making up living cells, we find that about 96 % of those elements consist of four basic elements: oxygen, carbon, hydrogen and nitrogen. Most of the oxygen and hydrogen are found in the form of water (H_2O); other than that carbon is the dominating element. The reason for this is the unusual capability of the carbon atom to bind other atoms together. Carbon has the ability to form everything from very simple to very complicated molecules. We call such carbon compounds organic molecules. A simple example is methane, which consists of one carbon atom and four hydrogen atoms. Carbon is evidently the starting point for all life types found on the Earth. Could there be some other atom which would also be suitable as a building block for life? The only element that in any way reasonably resembles the chemical flexibility of carbon is silicon. But silicon creates bindings with other elements that are chemically weaker than does carbon. Additionally, it is not available in gaseous form in the same way that carbon is (in the form of carbon dioxide) and is therefore less mobile. Although silicon is much more common than carbon in the Earth's crust, life is nevertheless carbon based.

The Double Helix

Nowhere is the organisation and complexity of life so well represented as in the DNA molecule. Every species has within its cells this remarkable molecule that contains a comprehensive architecture, a hereditary code that determines how a complete organism of the species is constructed and will look. Through cell division (mitosis), the code can be replicated and reproduced billions and billions of times, creating every cell type necessary to build up a full-grown individual.

Figure 3.4 displays an overview showing the coupling between individual atoms and how they bind together to build up a DNA molecule. The double helix is kept together by transversal bases. A base consists of four different basic jigsaw pieces: molecules of the substances adenine, guanine, thymine and cytosine. In the figure, two such bases are shown in detail. Also illustrated is the fact that it is quite ordinary atoms that are the central building blocks of life.

The code is stored in the form of extremely long DNA molecules, often several hundred million bases long. They lie entangled in a complex way inside the chromosomes. If they could be unfolded into a straight line it would extend several centimetres long. Humans have 46 chromosomes and the total genome is estimated to contain about two billion bases. Researchers have shown that only slightly more than a percent of the available DNA actually has any function; the rest seems to be junk or remnants from

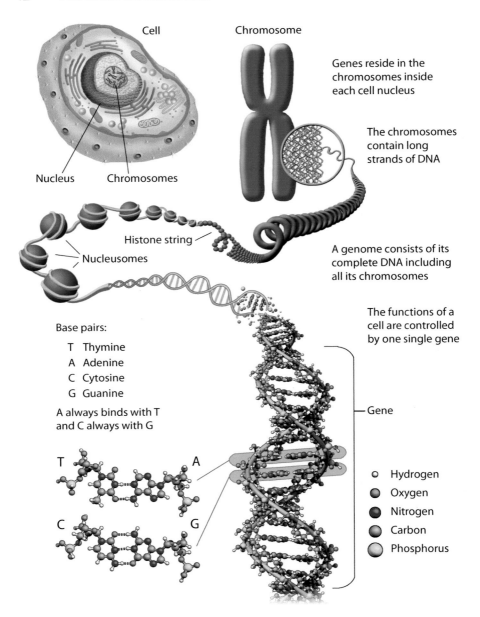

Fig. 3.4 The connection between the genome and the DNA molecules. Top: The genome is the sum of the chromosomes that exist in every cell. Every chromosome contains long threads of DNA molecules. Bottom: Part of a DNA molecule. Transversal nitrogenous base pairs are connected to a double helix backbone (marked by orange). Two examples of base pairs are shown in detail from above. The various atomic basic elements are colour marked (Eva Dagnegård)

an earlier evolution, but the study of this continues. The DNA molecule is partitioned into sections called genes. Every gene contains about 100,000 bases and humans have about 23,000 genes. The genes are the active components in the functioning of the DNA. There, the instructions for how to build various proteins are kept. The programming in the genes controls all necessary basic processes in an organism.

All life on Earth is DNA-based. DNA is found everywhere, in all beings, plants and animals down to microscopic life in the form of primitive single-cellular organisms. Different species have different DNA. Simple microbes have a simpler DNA, perhaps containing only a couple of hundred genes. On the other hand, there are species with a genome more complicated than humans'. As an example, rice has 37,000 genes. The modern research on DNA shows that all species have a lot in common and are quite similar to each other. It gives a powerful indication that all earthly life has a single and common origin.

How Was Life Created?

So far researchers have not succeeded in finding out how life was created. Part of the problem is to define the difference between life and non-life, which often is difficult in the microscopic world. In modern times several experiments have been carried out, simulating conditions which are thought to have prevailed when the Earth was very young and then trying to create primitive life. So far, the success has been limited. Researchers have been able to create several of the building blocks of life in the form of complicated organic molecules, but some steps are missing before you can call it life. As the American astronomer Carl Sagan expressed it: the musical notes are there but the music itself is missing.

The necessary basic material for life was surely widely available on primeval Earth. Hot, mineral-rich material was released from the ocean floors, combining with water to create organic compounds by means of chemical reactions. Simultaneously, organic molecules came to Earth contained in the continuous downfall of meteorites from space. Somewhere, a "primordial soup" existed where the conditions have been especially good for chemistry to transform into biology. The borderline between the two phases is very difficult to nail down. We know that there was a predecessor to the extremely complicated DNA molecule that we have already discussed, because it still exists. It is called RNA. RNA has a single helix structure and shows some of the characteristics that are typical for DNA, among them the capability to transfer genetic information, i.e. create copies of itself. The very first fragments of RNA molecules may have been formed from the organic compounds that were available in the "primordial soup". However, this cannot have happened out of pure chance—suitable minerals ought to have been present in order to increase the reaction speed. A simple form of membrane around the RNA developed, leading to the first stages of a primitive cell. This protection subsequently contributed to further develop the RNA molecules through mutations, becoming successively more intricate until it eventually had the capacity to copy itself. Already at this molecular level, the Darwin principle played an important role since more stable and copy-capable RNA molecules began to spread more and more. Eventually, the much more complex DNA molecule was formed, which today carries the genome for all living beings.

The Role of Oxygen

Clearly, any traces of what happened when life was created are very weak and difficult to interpret since so much time has passed since it occurred. The first traces seem to be about 3.5 billion years old. In Australia and Africa, microfossils of what is assumed to be primitive cells have been found. It seems clear that life started very early in Earth's history, perhaps as early as 4 billion years ago. Today, at least one species has arisen to achieve self-awareness becoming capable of effectively store and transfer information to the next generation. This has led to a cultural development and today's highly technological civilisation. But what actually happened along that road? Has the evolution theory of Darwin calmly, peacefully and inevitably advanced life into an intelligent human species?

No, it is not that simple, by far. The tracks that the scientists find from the Earth's past tell a different story. Sometimes evolution was at a standstill, at other times it progressed at great speed. Further, many exterior and random factors influenced its development.

The original atmosphere of the Earth had no oxygen. This means that the first primitive life must have been anaerobic—it did not use oxygen for its life functions, but instead received necessary carbon from carbon dioxide dissolved in the oceans and energy from non-organic chemical reactions. Such life today is considered to be exotic, but it still exists, for example at hot sulphur-rich underwater outlets in the ocean floors (Fig. 3.5). We call some of these life forms extremophiles.

The role of atmospheric oxygen in the development of life is in itself an exciting chapter. It is certainly true that the Earth's current high oxygen level is maintained by photosynthesis of plant life, but photosynthesis did not play a major role until relatively late in the evolution of the Earth. Before then, Earth must have had a different source for oxygen. Scientists believe that the same kind of organisms that are today responsible for algal blooms every

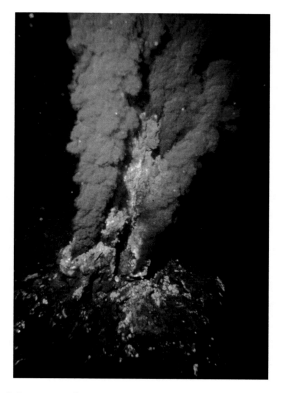

FIG. 3.5 Alien life on Earth? Anaerobic extremophiles, which do not need neither light nor oxygen, are thriving in hot hydrothermal vents on the ocean floors (OAR/NURP/NOAA)

summer, were responsible for the gradual increase of oxygen in the Earth's atmosphere about three billion years ago. They are sometimes called blue-green algae but are in reality not algae at all. Rather, they are primitive organisms, called cyanobacteria. It took about two billion years to reach modern levels of oxygen content, but without the oxygen production of these organisms, life would never had reached advanced levels. The oxygen content in the atmosphere of the Earth reached even higher levels when life became sufficiently complex to start photosynthesis. This probably happened about 2.5 billion years ago, and meant a decisive threshold in the progress of evolution. Oxygen is a chemically reactive element, and for many of the then existing life forms it became a deadly poison. Many types of microbes must have perished, but probably the increase of oxygen content was so gradual that others had time to adapt.

The mechanism for the development of oxygen on Earth is important when you consider the possibilities for alien life. Was the existence of cyanobacteria a fortunate random event that, after four billion years, finally allowed the formation of life at a higher level than microscopic life? Or was it possibly the other way around, that some factor delayed the evolution on Earth and that other planets may have developed higher life forms in a much shorter time? We do not have the answer to this important question.

Another very important step forwards for life was that the cell itself developed its own nucleus, in contrast to, for example, bacteria. We call such nucleus-bearing cells eukaryotes. Cells of this type started to appear about 2.1 billion years ago and were probably the result of an early symbiosis between primitive cells and small bacteria. Eukaryotes were news with a great potential for further development, which eventually would lead to plants and animals. However, until a billion years ago, all life on Earth was microscopic. From then on, multi-cellular organisms made slow and steady progress.

Then something dramatic happened. About 540 million years ago the development of life seems to have exploded in all directions at the same time. We call this the Cambrian period. There is much evidence that practically all basic forms of higher developed life emerged during this intensive period, which only lasted for about 40 million years, geologically a very short time. Paleontologists have not clarified what happened in detail, nor why. Perhaps the oxygen level was now sufficiently high for higher forms of organisms. Perhaps the DNA of the eukaryotes now had reached sufficient complexity that plenty of new species could be formed in a very short time. Perhaps the climate changed in a critical way from cold to warm at the same time as the initial absence of predators allowed improved survival conditions for more individuals. At any rate, we and most other animals can track our roots back to this active period. But there are scientists that believe that the "explosion" was not all that abrupt and that some new species had developed already before the beginning of this era.

The next major step for life was the climb to dry land. The first plants seem to have taken this step about 475 million years ago. A precondition for this was, however, that harmful radiation from space had decreased to endurable levels. When the level of oxygen gas had reached a sufficient quantity, an ozone layer gradually started to build up. Ozone consists of three oxygen atoms, O_3, compared to the two of oxygen gas (O_2) and is created through the influence on oxygen gas by ultraviolet radiation from the Sun. Ozone shields and protects the surface (together with the magnetic field) from the solar wind and cosmic radiation.

The original microbe life ought to have been present in small ponds on land, and these may have constituted a basis for the conquest of land. Less than half a billion years ago, plants started to proliferate on land, and then it was time for animal life to develop, since there now existed plants to live from. About 360 million years ago, there were already giant forest full of insects, forests that, by the way, eventually were transformed into the coal that became the foundation for the industrial revolution in much more recent times.

Extensive Extinction

The evidence in favour of Darwin's theory of evolution is overwhelming. At the same time, it is obvious that evolution progressed irregularly, and that changes in conditions and environments created both opportunities and limitations. What Darwin could not have known was that life has been subjected—several times—to great strain in the form of mass extinction and global catastrophes.

Fig. 3.6 A gigantic collision 65 million years ago meant the end for the dinosaurs—and a possibility for the mammals (Mike Carroll)

The most well known case happened 65 million years ago (Fig. 3.6). At this time, a massive celestial body hit the Yucatan peninsula in Mexico. The impactor was most probably an asteroid with a diameter of 10–15 km. The catastrophe was enormous. The energy that developed can be compared to the explosive power of a hundred million hydrogen bombs. The strike partly hit in the sea, generating enormous tsunami waves. The impact threw vast amounts of dust into the atmosphere, and resulted in wide-spread forest fires.

All life close to the impact point was immediately erased. But far worse was the dust and smoke filling the air, gradually spreading on global scale. This cloud blocked the sunlight and caused a long period of low temperatures. The vegetation was heavily reduced, affecting many animal species. It is now generally accepted that this event caused the downfall of the dinosaurs, which until then had dominated the Earth for 135 million years. Marine life was also affected, for instance by poisonous gases dissolving into the water. But life recovered anew. Mammals now had a chance to appear on the scene, previously effectively secluded by the dominance of the dinosaurs.

The asteroid collision scenario is supported by strong evidence. Traces from the impact have been found in the form of a 200 km sized crater ring close to the Mexican coastline. The global consequences of the catastrophe can be seen in sediments all over the world. In particular, it has been shown that the element iridium is over-represented in layers from this epoch. Iridium is a rare element on Earth but quite common in asteroids—so the iridium content in the layers should have come from the asteroid impact. A well-known example of these sediments is found at Stevns Klint south of Copenhagen in Denmark.

The catastrophe 65 million years ago is far from the only one that has taken place in recorded history. At least four other mass extinction events have been identified over the last 500 million years. In Fig. 3.7 we see a diagram marking these, the asteroid impact is E. The other events may have had similar or other causes, perhaps in combination with each other. Very strong and prolonged volcanic eruptions may be an explanation, where hundreds of thousands of square kilometres may have been covered with lava, simultaneously polluting the atmosphere. The amount of atmospheric blocking of dangerous radiation from space may have varied, which in turn may have killed many species and additionally

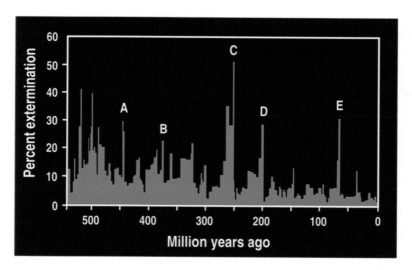

Fig. 3.7 The five large mass extinction events during the latest 500 million years. The percentage scale concerns fraction of whole families of exterminated species (Peter Linde)

led to frequent mutations. Another possibility is that Earth may have been subjected to sterilisation from supernova explosions in our neighbourhood. A star that annihilates itself in such a cataclysmic way transmits enormous amounts of dangerous radiation.

The story of life's continuous evolution during the last 50 million years is full of exciting elements, but is unfortunately beyond our scope for the moment. It should, however, be pointed out that the human part of Earth's history is, geologically speaking, infinitely small. The early forefathers of the today's apes, as well as of humankind, lived about 10 million years ago. Modern man appeared about 100,000 years ago. If you compress Earth's history into a 24-h day, this corresponds to little more than a 1 second tick. Our high-technology civilisation began only a hundred years ago, which would correspond to the latest millisecond. Let us hope that it will not be the last, and look forward to the next...

Resiliency of Life

Evolution has pushed, or possibly encouraged, life to occupy all niches in the environment. Life forms exist that not only survive but even thrive in conditions you could hardly envision as life friendly. Such forms are called extremophiles, and are normally microbes. We have already encountered these organisms, which live without light or oxygen at hot springs on the floor of the oceans. Some extremophiles can survive temperatures in excess of 130 °C, while others manage strong cold close to absolute zero. A somewhat larger being is the tardigrade (also known as the water bear, (Fig. 3.8), which can grow to a size of about 1 mm. It has been exposed to a variety of very strenuous experiments showing its incredible endurance. By lowering its metabolism to a ten thousandth of normal, and its water content to 1 %, it can survive extreme temperatures as well as harsh radiation conditions and large pressure gradients. Space experiments, for example, have shown that it survives vacuum for weeks.

Extremophiles can sometimes feed themselves in unexpected ways. On board the International Space Station (ISS) a certain fauna of bacteria and microbes have developed. Some of them thrive on plastic materials and others corrode metals like aluminium and magnesium.

FIG. 3.8 Water bears can endure the most varying conditions. The picture is taken with a scanning electron microscope (Eye of Science/Science Source)

How Much Is Pure Chance?

It seems probable that life emerges relatively easily if circumstances are right. On Earth this evidently happened quite soon after the stabilisation of the newborn planet. But during billions of years, life was still microscopic and primitive. We see that evolution has been affected in many ways by random events that have driven it in many directions. There are many difficult thresholds that life must have overcome in order to arrive at the level of today. Examples are the formation of the DNA and of cells with nuclei (eukaryotes). On the other hand, life—at least on microbial level—seems to be capable of surviving extreme conditions, perhaps even more extreme than we find on other planets in the solar system. However, one thing is surviving in such environments, another is whether life can arise in such difficult conditions.

The decisive question is really whether the emergence of mankind and intelligence in any way was inevitable. Unfortunately, we don't know the answer. We can speculate on what would have happened if the dinosaurs did not receive a death blow by an asteroid. Would we ourselves still be reptiles and not mammals? The dinosaurs dominated the Earth for about 135 million years, an incredible length of time, especially compared to the existence of mankind.

They were active and adaptable animals, but intelligence does not seem to have been the most important survival factor. Perhaps the world would be much the same today as it was then if the dinosaurs had not become extinct. This book would certainly not have been written and there would most probably be no one on Earth to ponder the mysteries of the universe.

So is the human DNA something absolutely unique? Actually not, but clearly a very small difference in the genetic code may have important consequences. Man shares 98 % of its DNA with modern apes but is nevertheless the only species that has developed modern technology, and the only species who is studying its origin and planning for its future.

Might an evolution similar to the one that happened on Earth have happened on other planets? We will return to that question in a little while.

4. Alien Life at Close Range?

The possibility of finding intelligent life on other worlds has fascinated mankind for centuries. With the introduction of the telescope in the seventeenth century, the hypothesis became more concrete, especially in the realms of our own solar system. With the telescope, it was possible to see that the planets were not merely points of light, under magnification they actually showed disks. Observers realised rather soon that Venus and Mars were the most interesting ones. They are our nearest neighbours; relative to the Sun, Venus is orbiting inside and Mars outside. The size of Venus could be estimated, and it turned out to be much the same as Earth's. In addition, it seemed to have an atmosphere, but this was unfortunately so dense that it remained impossible to see the actual surface of the planet.

Any Monsters Nearby?

It took until the beginning of the 1960s before a realistic understanding of the conditions on Venus became possible. Before then it was free to speculate about the possibility of intelligent Venusians. Many believed that Venus had an acceptable climate—albeit quite warm. Svante Arrhenius, Swedish Nobel prize winner, suggested in 1918 that Venus—to a large extent—would consist of swamps. Science fiction literature heavily exploited this "fact" during the 1930–1940s. Many monsters had their origin in the Venusian swamps. Perhaps this contributed to the belief amongst the common man that alien beings existed—although it would be wise to avoid them.

However, if you wanted to meet aliens, the planet Mars seemed to be the most probable choice. During the nineteenth century, scientists aimed bigger and bigger telescopes at the planet. They discerned more and more details on the Martian surface,

© Springer International Publishing Switzerland 2016
P. Linde, *The Hunt for Alien Life*, Astronomers' Universe,
DOI 10.1007/978-3-319-24118-0_4

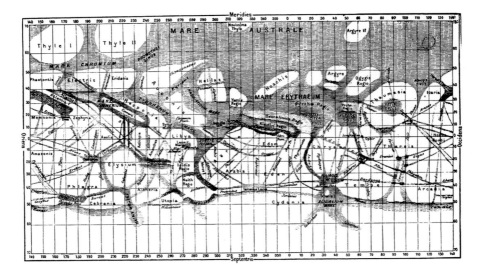

FIG. 4.1 The map of Schiaparelli from 1877, showing the planet Mars and its alleged canals

including clear evidence for polar caps strongly reminiscent of the ones on Earth. Just like Earth, Mars is inclined in its orbit, which causes seasonal changes. And indeed, the Martian surface seemed to change characteristics with the time of year. The polar caps also changed with the seasons. At the end of the nineteenth century, speculations about Martian civilisations began to flourish. In 1877 the Italian astronomer Giovanni Schiaparelli aimed his large telescope towards Mars, which then was closer to Earth than it had been for a long time. He saw distinct formations on the surface that he called "seas" and "continents". Additionally, he claimed to see narrow streaks connecting the continents with each other (see Fig. 4.1). These he called "canali", which, erroneously, was translated as "canals". The word "canali" may also mean groove or channel, not necessarily implying an artificial origin. However, the term "canals" leads the mind towards artificial constructions, and supported speculations about an advanced Martian civilisation.

The wealthy American Percival Lowell was convinced that the canals existed. He erected an advanced private observatory to study them in detail. At the end of the nineteenth century he published a large number of observations of the alleged Martian canals, and in 1906 Lowell explained in a grand and glorifying New York Times interview that intelligent life on Mars was a proven fact. At the same time, the famous physicist Nikola Tesla declared that he had built a radio receiver that had received signals from Mars. In addition to this came the publication of the famous novel *The War of the Worlds*, written by the English author H.G. Wells. In this, he describes how a superior but evil Martian civilisation decides to invade and occupy the Earth. In the eyes of the general public, it was now commonly accepted that Martians existed. However, more serious astronomers were sceptical.

Mars Attacks!

In 1938 something happened that in other circumstances could have been considered one of the most successful April fools jokes ever conceived. However, it was not meant as a joke, nor did it happen on April 1. On October 30 of 1938, a dramatic radio theatre play called *The War of the Worlds* was broadcast all over the USA. Estimates hold that it reached several million listeners. For the responsible producer, a hitherto unknown director and actor with the name Orson Welles, this night would turn out to be quite tumultuous. Welles had moved the original plot from England to New Jersey in the USA. This heightened its credibility, especially for the many who did not hear the programme from the beginning. Welles presented the radio broadcast's course of events in the form of dramatic bulletin news that every so often interrupted an innocent entertainment programme. In the play, the Martians rapidly advance against New York, and in one scene mass deaths are described by a reporter who himself succumbs. The reaction from the listeners was strong, and in some places panic broke out. Many were ready to evacuate, the police and other authorities were overwhelmed by worried phone calls. Although the authorities did their best to calm down the situation, rumours about the invasion spread with lightning speed.

In the end, the general public calmed down. Orson Welles had drastically underestimated the effect of the dramatisation when he commented that "it was our thought that perhaps people might be bored or annoyed at hearing a tale so improbable". He was wrong;

the following month more than 12,000 newspaper articles were written covering the story.

That the impact of this dramatised Mars invasion became so large clearly showed that the general public was ready to believe in the existence of Martians. It should be noted that this happened in a dramatic age; the public was already on edge. The year before the airship Hindenburg burned in a colossal catastrophe that was made immortal by a dramatic live broadcast on radio, and Orson Welles used this as a model in his drama. Additionally, world politics at the time was very tense, as an aggressive Hitler had just began his military conquests. Admittedly it must be said that the dramatisation was cleverly done—even a modern listener can easily be drawn into the dramatic flow of events. Today the broadcast can be relived and enjoyed on Internet in its original form. Orson Welles later went on to become a world famous actor and director.

Percival Lowell and his pet theory about Mars eventually lost its credibility. Mars was not a dying planet inhabited by a civilisation trying to save its planet with an intricate system of artificial irrigation canals. On the contrary, in 1965 the notorious canals were dissolved into thin air when the spacecraft Mariner 4 arrived at the planet for the first close-up pictures. No canals were seen; they proved to have been nothing more than optical illusion. Instead, a world appeared that in many ways seemed more similar to the Moon than to the Earth—dry, lifeless and filled with craters. However, modern research shows that Mars must have had a completely different climate during its early existence. The hope is still alive about finding life on Mars.

Modern Research

Today, the seriously-minded search for life in the universe still begins at close range, in our own solar system. We have superficially acquainted ourselves with the planetary system in the previous chapter. Now, let us have look at whether alien life can be found close to the Earth.

As our starting point we assume in this chapter—as in most of this book—that with life we mean something that looks like terrestrial biological life. We should be able to recognise it as life, with the typical characteristics we previously discussed. This might be considered a rather narrow perspective. However, to distance ourselves too far from this view would rapidly lead us into quagmires of speculations, where our current understanding is too limited to lead to a fruitful discussion. Nevertheless, we will eventually return to a few rather exotic ideas about how life is supposed to be created and further developed.

The Habitable Zone

A common—and reasonable—assumption is that life and water have a close link. We are rather sure that life on Earth was initiated in the sea, in which case we have an obvious clue to where we should be looking for life in the first place. We want to find water, and we want it to be in liquid form. This already sets limits to where reasonable searches should be made, searches that bring us to a region we call the *habitable zone*. We will use this concept quite a lot in our discussion about the newly discovered exoplanets. So let us take a closer look at what it means.

The simplest and most general definition of the habitable zone is the distance from a star within which its energy is just sufficient to keep the water on a planet in liquid form. It should not be too cold so that the water freezes into eternal ice, and it should not be too hot, transforming all water into vapour. Knowing that planetary orbits may be quite elliptical, it is an important additional requirement that the orbit of the planet is sufficiently circular so that it stays inside the zone. Since there are many types of stars, both larger and smaller than our Sun, their energy output are also different. However, for most stable stars one can speak of a habitable zone. An important group is the red dwarfs, stars that are smaller and cooler than our Sun. Their habitable zones lie closer to the host star and are also more narrow.

There is a habitable zone in our own solar system; the Earth of course is the very proof of this. In Fig. 4.2 we see the zone marked in green. Scientists, however, vary in their beliefs for the actual boundaries of this zone. In the figure, the light green area corresponds to the most generous definition, while the inner, dark green, corresponds to the most conservative. Of course, the Earth

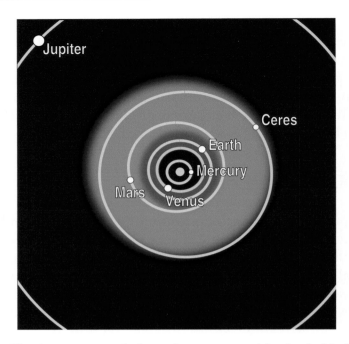

FIG. 4.2 The inner parts of the solar system with the habitable zone marked. The *light green* colour references a liberal estimation while the *dark green* corresponds to a more conservative estimation (Peter Linde)

orbit is part of both alternatives. But are there more planets in the solar system that can be considered to be inside the zone? The question actually is not straightforward to answer since there are plenty of complications when it comes to the perception of habitability. The zone can be both larger and smaller depending on the actual conditions of the planet. In fact, as we will see, habitable environments may be found in quite unexpected places.

Atmosphere Is Important

Perhaps the dominant factor is whether the planet has an atmosphere. In the vacuum of space, water can only exist either in the form of vapour or ice, not as a liquid. Even if the planet is located well within the habitable zone, a suitable atmospheric pressure is necessary for water to exist as a liquid. Also in many other ways the presence of an atmosphere affects conditions on a planet.

A greenhouse effect increases the temperature higher than it otherwise would be. On the other hand, strong cloud production leads to an increased *albedo*, which means that a fraction of the Sun's light is directly reflected away before it reaches the surface (see Fig. 3.2). This lowers the surface temperature. In addition, a reasonably dense atmosphere strongly assists in distributing heat over the entire planet, thereby equalising the temperature. There may well be planets in other planetary systems where this mechanism is a decisive factor for sustaining life. The reason is because *bound rotation* may be commonplace. Bound rotation is a kind of gravitational phenomenon in which bodies that lie in orbit close to a larger object are locked to always show the same side towards the larger object. This means that the smaller object rotates one rotation around its own axis simultaneously as it orbits once around the larger object. The Earth-Moon system is an example of this phenomenon. For a planet with bound rotation with respect to its star this would mean that the front side receives all the energy and the backside nothing. But, as mentioned, the existence of an atmosphere can balance this temperature difference to some extent.

Whether a planet has an atmosphere or not depends on many factors, in particular on its gravitation and its temperature. Gravitation determines the *escape velocity*, the minimal necessary velocity an object needs in order to break free from the surface of a planet and enter into space. In the case of the Earth, the escape velocity is 11.2 km/s or about 40,000 km/h. To be clear, this applies to ballistic bodies, meaning objects that—after gaining their initial speed—do not have any further power source (for example a cannon projectile). In the 1865 Jules Verne novel *From the Earth to the Moon*, three daring men ride inside an enormous cannonball blasted away with a gigantic cannon. It would never work in practice. Neglecting the little detail that passengers would immediately be annihilated by the acceleration, a cannon projectile can never reach the necessary speed. The limit is set by the propagation speed in air of the blast's shock wave. In heavy guns an exit velocity of about 1 km/s can be reached, which clearly is far from enough. But by using hydrogen or helium as medium instead of air, special cannons have reached exit velocities of about 7 km/s. Even this velocity is insufficient for escaping the grip of the Earth.

Currently, the only way to escape from the Earth's gravitational field is by means of rocket propulsion.

Other bodies have other escape velocities, depending on mass, density, etc. As a comparison, the escape velocity of Mars is 5 km/s, the Moon's is 2.4 km/s and Jupiter's is 60 km/s.

The escape velocity is also an important factor in the microscopic world. Which speeds are typical for the molecules in a normal gas? What we in the macroscopic world perceive as heat is perceived in the microscopic world as merely the motion of molecules. Some basic rules apply. At a given temperature, the lighter molecules in a gas are moving faster than the heavier ones. At a higher temperature all molecules move faster than at a lower temperature. It is possible to calculate the speed of the individual gas molecules, and it turns out that they are rather high. At room temperature oxygen molecules move at a speed of about 0.4 km/s. The corresponding value for hydrogen gas is 2 km/s. At 1000 °C, hydrogen is moving at 4 km/s. These are average values, but the motion of molecules follows a statistical distribution, meaning that some are considerably faster than the average. The consequence may be that individual gas molecules sometimes exceed the escape velocity and spontaneously disappears from the atmosphere of a planet out into space. So, from these considerations, we see that high temperature and low gravitation aggravate the existence of a planetary atmosphere.

Liquid Water

There are more complications concerning the habitable zone. Not all planets are in a completely circular orbit around their stars. In our solar system Mercury is the most pronounced example of this. As we shall see in due course, strongly elliptical orbits are commonplace among exoplanets. An elliptical orbit implies that a planet receives varying amounts of radiation from its parent star which may have the consequence that liquid water can be boiled off at the position closest to the star, and freeze to ice further away. Other factors may also play a role for water to remain in liquid state. Oceans on Earth consist of salt water; and this somewhat lowers the risk for freezing, since the freezing point for salt water is about −2 °C. A mixture with other substances, for instance ammonia, may further lower the freezing point. Another extremely important and unique property of water is, as previously mentioned, that the density of water decreases when it solidifies, i.e. transforms into ice. So, ice floats in water instead of sinks, which strongly reduces the risk for deep-freezing an entire lake.

The size of the habitable zone can vary for many reasons. A quick look at Fig. 4.2 shows that Mercury is too close and that Jupiter is too far away. But is that entirely true? As we will soon see, there are other energy sources than the Sun in our solar system, which may alter our scenario somewhat.

Space Exploration in the Solar System

We have a distinct advantage concerning the exploration of our own solar system: we can study it at close range. With the largest Earth-based telescopes it is possible to arrive at some preliminary conclusions. But even more important is the fact that our solar system has come within travel distance over the past 50 years. We now have the possibility to find out *in situ*, on site, the conditions on other planets in our neighbourhood. Unfortunately, similar exploration flights to the newly discovered exoplanets will not become possible until far into the future, if ever (but see Chap. 12).

The first time a vehicle made by man came into contact with another celestial body was in 1959, when the Soviet spacecraft Luna 2 crashed on the Lunar surface. Soon after, Luna 3 succeeded in taking the first images of the backside of the Moon. A space race between the USA and the Soviet Union had begun, fueled by a political and ideological power struggle. It had the positive side effect that the research of the solar system continued at a high intensity. Manned spaceflight certainly had a greater propagandistic value, but until the fall of the Soviet system in 1990, a number of very successful unmanned exploration flights were undertaken. Thereafter, primarily the USA (NASA), Europe (ESA) and Japan have contributed to the exploration of the solar system, although new nations like India and China are beginning space explorations. All the planets have been visited by space probes from Earth. In addition, a number of moons, asteroids and comets have been investigated. Even Pluto is in the process of being studied. The spacecraft New Horizons was sent to Pluto in 2006 and arrived in July 2015. Initial data shows a surprisingly young surface possibly affected both from surface–atmosphere interactions as well as impact, tectonic and cryovolcanic processes. Shortly after its launch the International Astronomical Union decided to degrade Pluto's status from being a planet to being a dwarf planet...

Mercury: Fried by the Sun

As seen from the Earth, Mercury is a difficult planet to observe, even if occasionally it can be seen by naked eye close to the Sun in the evening or morning sky. A more unusual way of observing Mercury is in silhouette when it sometimes passes in front of the Sun. This happens at uneven intervals; the next Mercury passages that can be seen from Earth take place on May 9, 2016 and Nov 11, 2019.

Mercury is the planet closest to the Sun and only needs 88 days to make one orbit around the Sun. It has a rather elliptical orbit, and its average distance to the Sun is 0.39 AU. This results in Mercury receiving between five and ten times more solar energy than the Earth. It is a rocky planet, about a quarter the size of Earth. In 1974 the space probe Mariner 10 was sent there to observe the planet at close range. The pictures transmitted back immediately made it clear that the planet was completely covered with craters, much the same as our own Moon. Not surprisingly, it also turned out to be very hot. The highest temperature at the equator of the sunny side was measured at close to 450 °C. It was also no surprise that no signs of an atmosphere were found. The small size of Mercury combined with the high temperature means that it cannot be expected hold any atmosphere.

The planet also has another peculiarity. Since it is located so close to the Sun, astronomers long believed that the planet always showed the same side towards the Sun, i.e. that its rotation was bound. However, closer studies have shown a somewhat more complicated relationship, namely that the planet rotates three times around its own axis at the same time that it orbits the Sun twice. This a gravitational resonance of a more unusual kind. In practice this means that the whole of Mercury is subjected to both intense heat and strong cold. There is, though, one zone of exception, another interesting peculiarity: Mercury's rotational axis is barely inclined with respect to its orbital plane, just 2°. For comparison, the inclination of the Earth's axis is 23.5° with respect to its orbital plane, which causes the seasonal changes on Earth. In the case of Mercury, the small inclination means that there is a zone at its Southern and Northern poles where at the bottoms of the craters neither heat nor cold can dominate. From these places radar and other studies have indicated the existence of what is believed to be water ice.

In 2004 a new space probe, Messenger, was sent to Mercury. After a long series of gravity assist manoeuvres and fly-bys, it went into orbit in 2011. Gravity assist is a way to gain or loose speed by passing near a main body, like a planet. This is achieved by moving close to the gravitational field of the planet and taking advantage of its motion. To gain speed it works like a kind of slingshot effect, which hurls the craft away. But it can likewise be utilised to decrease the speed, in this case necessary to reach Mercury. Since 2011 thousands of close-ups have been sent to Earth. One of them is seen in Fig. 4.3. After more than 4000 orbits, Messenger ended its life in 2015 by crashing on the planet, after several extensions of its highly successful mission. In 2017 another expedition, called

FIG. 4.3 The surface of Mercury imaged at close range. The larger crater is Strindberg, named after Swedish author August Strindberg (NASA)

BepiColombo, is expected to depart for Mercury. This project is a collaboration between Europe and Japan. The plan calls for the joint launch of two spacecraft. The journey will take considerable time, again using several gravity assists to decelerate, and BepiColombo is not expected to arrive until 2024, when it will enter into orbit around the planet.

The chances of finding life on Mercury must be considered negligible. Water in liquid form has most probably never been available. Future space expeditions may of course land in the polar regions and take a closer look. But there are other places in the solar system with higher priority.

Venus: Hotter Than a Baking Oven

A closer look at Fig. 4.2 shows that Venus lies marginally within the innermost vicinity of the habitable zone of our solar system. Occasionally, it may come quite close to Earth and is then seen as a very bright object in the sky. In fact, Venus is the brightest object than we can see in our sky, after the Sun and the Moon. In early times, many speculated that Venus could be very similar to Earth. As we mentioned earlier, it is almost the same size. It was suspected to have a very dense atmosphere, because even the largest telescopes never revealed any details of its surface. Observations soon confirmed that the atmosphere of Venus simply was too dense to allow any surface viewing.

For a long time, the lack of information provided fertile ground for all kinds of speculative suggestions regarding life on Venus. Not until the 1960s more conclusive evidence was presented. Radar-based investigations from Earth returned echoes, and from those it became possible to extract some information. At the same time it became feasible to send space probes there. The Soviets excelled in this. Nevertheless, it was the American Mariner 2 spacecraft that was first on the scene to make measurements. It turned out that the surface temperature was unexpectedly high, about 400 °C, a great disappointment for the optimists that still hoped for Venusian life. Among them were many science fiction writers who subsequently had to leave the solar system in their continued adventures with aliens. As we will soon see, at the same time Mars also turned out to be very inhospitable for monsters.

Since then about 20 space probes have been sent to Venus and a further dozen or so have passed nearby in their quest for other targets. During more than 20 years the Soviet Union sent a number of space probes to Venus with the name Venera. In 1967 Venera 4 attempted to reach the surface using a parachute for braking, but the atmosphere was so dense that the falling time became unexpectedly long. Unfortunately, this meant that the batteries of the probe were exhausted before the actual landing, but successful readings of the atmosphere were taken, showing that it consisted mainly of carbon dioxide.

Learning from this experience, engineers redesigned later Venera probes, and at several occasions they finally succeeded in landing and transmitting images from the Venusian ground. Soviet space research always had its roots more in raw strength than in technological finesse, and resilience was needed for the Venus landers. The atmospheric pressure turned out to be more than 90 times higher than the Earth's. This is an extreme pressure, comparable to the pressure found on Earth in an ocean at a depth of 1000 m. The temperature was also extreme, 465 °C. The scientists had planned for a half hour of survival time but the lander actually held out for two hours. That was enough time to allow for sending valuable images and other data to Earth. These pictures are still the only ones that have been taken of the Venusian surface and they represent, in many ways, an extraordinary achievement.

In 1990 the NASA space probe Magellan arrived at Venus. During almost five years, it completed thousands of orbits around the planet. The craft was equipped with an advanced radar which continuously scanned the Venusian surface, right through the dense atmosphere. The craft transmitted its results back to Earth and, by means of sophisticated image processing techniques, the data was transformed into detailed three-dimensional images of the surface. Figure 4.4 shows one example. At last, the astronomers had gained a clear view of what the Venusian surface looked like.

The surface is rather different from that of other planets. Although there are a number of impact craters, more typical are large volcanic plains. Almost 200 very large volcanoes have been identified. Images suggest that recent eruptions from them have

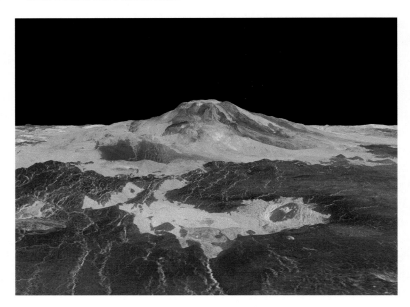

Fɪɢ. **4.4** The largest volcano on Venus, Matt Mons, is approximately 8 km high. The image is generated from the space based radar measurements made by the Magellan spacecraft. The height scale is exaggerated in order to emphasise the structures (NASA/JPL)

deposited today's dominant surface structures. There are signs that volcanic activity is still prevalent. Although no active volcano has been observed directly, the content of sulphur seems to have varied between measurements made at different times. Contrary to the situation on Mars, which we will soon take a closer look at, there are no apparent traces of flowing liquids from an earlier epoch. Venus has a surprisingly long rotation time around its axis, no less than 243 days, and is—in addition—rotating in retrograde direction. This strange fact may have resulted from violent impacts in the early history of the planet. Together with the absence of plate tectonics, this would explain why Venus, unlike the Earth, does not posses a strong magnetic field to protect it from solar wind and cosmic radiation.

Conditions on Venus are quite extreme. Violent thunderstorms have been registered by a couple of spacecraft. The temperature can rise to almost 500 °C, enough to melt metals like tin, lead and zinc. The atmosphere is exceedingly dense and contains no oxygen. Its density distributes the high temperatures all over Venus, on the daylight side as well as the night side. In the upper parts of the atmosphere, clouds consist mainly of sulphuric acid. This can fall as rain but probably never reaches the surfaces before being vaporised again.

What Happened to Venus?

How did the climate get so completely out of hand for a planet that is, albeit narrowly, inside the habitable zone? Researchers deduce that the extreme temperature is actually not due to the planet's proximity to the Sun. The amount of solar energy that reaches the surface of Venus is actually less than that reaching the Earth since the atmosphere reflects back into space most of the solar energy. Instead, a very strong greenhouse effect explains the high temperature. On Venus this effect has reached extreme levels, levels which may have been present a long time ago. From the beginning, Earth and Venus may have been quite similar. It is also possible that Venus may have had large quantities of water. Many scientists believe that the creation of the oceans of Earth has its origin in a heavy bombardment of large bodies, like asteroids or comets, mainly consisting of water ice. If that is the case, Venus would have received a corresponding amount. From then on, however, evolution of Venus and Earth has followed different tracks. It is conceivable that these early Venusian oceans simply boiled away in the era of the young solar system, when the Sun gradually increased its luminosity. The resulting water vapour worked as a very effective greenhouse gas which led to a strong rise in temperature. Things got even worse when rocky minerals began to release carbon dioxide. The level of this gas rapidly increased in the atmosphere, which further elevated the temperature to the current level. However, in the modern Venusian atmosphere most of the water vapour has disappeared or turned into other forms. The solar radiation has gradually separated the water molecules into its constituents, hydrogen and oxygen. Most of the hydrogen would have disappeared into space. If this theory is correct, the remnant hydrogen still present should show an increased fraction of deuterium, i.e. hydrogen nuclei with an extra neutron. Such atoms are thus heavier and should have more difficulty in escaping into space. Precisely this has been found by the space probes visiting

Venus. The remaining oxygen has partly diffused into space and partly become bound into carbon dioxide.

In spite of its superficial similarities to Earth, it is not an exaggeration to argue that Venus is the planet most hostile to life in the entire solar system. If a hell exists somewhere, some say that Venus would be the obvious place.

In spite of this, there are speculations about life on Venus, albeit not on the surface itself. It turns out that some distance up in the atmosphere, conditions are considerably better. At about 50 km altitude the temperature has dropped to 20 °C and at the same time the pressure has decreased to the same as at sea level on Earth. Paradoxically enough, this means that this zone in the Venusian atmosphere represents conditions closest to the Earth that can be found anywhere in our solar system. In principle, bacterial life could thrive there but no indications have been found that this is the case. However, there are speculations that future explorations of the planet could take advantage of these circumstances. Airships of the Zeppelin type could conceivably hover around for long periods of time in the atmosphere. Perhaps future colonies could be established on floating platforms located at a suitable height in the Venusian atmosphere. However, while waiting for such science fiction-inspired initiatives, the more mundane exploration of Venus goes on. Since 2006 the European spacecraft Venus Express has been orbiting the planet. It collected new data on the atmospheric and ionospheric conditions, including the discovery of a small ozone layer. In late 2014, the last remaining manoeuvering propellant was used to execute a series of aerobrake operations to analyse lower parts of the Venusian atmosphere.

Mars: The Hope Remains

The planet Mars has always been the object of speculations, some of them quite fanciful and far from the reality we now believe we know (Fig. 4.5). Mars is usually quite difficult to observe from Earth. It is an outer planet which means that its orbit around the Sun lies outside the Earth's orbit. Hence Mars receives less solar energy than the Earth. Additionally, the distance between Mars and Earth varies strongly, depending on where in their respective

FIG. 4.5 In 1893 Kirk wanted to sell soap to Martians. The business man Yerkes had just financed the worlds largest lens telescope

orbits they are located at the moment. It is, of course, when they are the closest that detailed observations can be made. This takes place when Mars is in *opposition*, i.e. in a straight line with the Sun and the Earth, which happens about every second year. Since the orbits of both Earth and Mars are elliptical, the distance between the two bodies at opposition will vary. During the 2014 opposition the distance was 93 million kilometres, while the next favourable opposition will be in 2018 when the distance is only 57.6 million kilometres. Even then, the apparent size of Mars in a telescope will not be much larger than a single average sized crater on the Moon.

It has always been a challenge to see details with ground based telescopes due to the blurring effect of the turbulence in the Earth's atmosphere. In spite of these difficulties, modern technology allows even amateur astronomers today to obtain excellent pictures of Mars. Using sensitive digital video imaging, many frames can be taken in a continuous mode, allowing the capture of those rare moments when a steady atmosphere permits clear images with high resolution. With a method known as "lucky imaging", all blurred images can be deleted and the remaining sharp images combined into a high resolution image.

The pictures from 1965 taken by the American Mariner 4 spacecraft effectively punctured all lingering romantic or horrifying notions of an advanced Martian civilisation. The crater-dominated surface surprised many.

But Mars actually displays some characteristics similar to the Earth's (Fig 4.6). Although it is only about half the size, it

FIG. 4.6 Mars observed with the Hubble telescope in 1999. The two polar caps are conspicuous. The dark areas are permanent formations. To the *right* is seen a cloud system above Olympus Mons, the largest volcano in the solar system (NASA/ESA/Hubble Heritage Team (STSci/AURA))

still is a rocky planet with permanent formations in different shades. For long very little was known about the nature of the atmosphere more than that it was transparent, contrary to the Venusian atmosphere—except when occasionally something seemed to block the view. This something was eventually identified as very heavy sand storms, sometimes covering the entire planet. Through observations of the surface features, it was possible to determine the length of a Martian day to 24 h and 37 min, just a little longer than Earth's. Likewise, the inclination of the planetary axis was measured to 25°; close to that of the Earth's, which is 23.5°. Evidently, Mars had seasons nearly identical to the Earth. For a long time, the white polar caps were also well known, resembling the ones on Earth. The sizes of the polar caps varied with the seasonal periods, which strongly indicated some kind of climate-dependent changes.

Intensive Research

A considerable number of spacecraft followed in the trail of Mariner 4. Without doubt, Mars is by far the most scrutinised planet of all. But with new knowledge, new questions have also arisen, and therefore the exploration will continue in the following decades and culminate with manned expeditions. Already, ideas are discussed about how to terraform Mars in order to make it more similar to the Earth. Perhaps in the distant future, a colonisation will then take place.

So far, almost fifty attempts have been made to reach Mars, either passing by, going into orbit or landing on the surface. Out of these, twenty have succeeded. Several of the successful attempts have resulted in unusual and spectacular data, such as detailed close-ups, analysis of Martian minerals and gas content, and much more. However, no canals have ever been discovered. Nevertheless the interest in Mars has remained high both among the general public and among scientists. That probably explains the fact that currently there are six different spacecraft around or on Mars that simultaneously are making detailed studies.

The reason for the great interest is obvious: there are certainly no Martians to be found, but life—or at least traces of earlier life—still cannot be ruled out. And this would be a revolutionary discovery in itself, since it would mean that life on Earth is not unique. Detailed studies are carried out along two main principles, one being a comprehensive mapping from space using orbiting probes, the other being direct landings in order to investigate the nature and composition of the ground material. In 1976, NASA achieved two successful and spectacular landings as the Viking 1 and the Viking 2 set down on the surface. Figure 4.7 shows the first historic picture of a real Martian landscape. What we see in the picture is a desert landscape which in many ways reminds us of similar places on Earth. The surface has a reddish colour, just like the majority of the planet. In fact, this causes the conspicuous red colour of the planet when observed by naked eye from Earth. The red colour is due to the surface layers being dominated by an iron compound similar to rust. It is probably a contributing reason to why it was named after the Roman war god Mars.

Since Viking, several successful landings have been made. Some landers have been mobile and have moved around on the planet, among them Sojourner 1997 and Spirit and Opportunity

Fɪɢ. 4.7 In 1976 Viking 1 revealed for the first time what a Martian landscape looks like (NASA/JPL-Caltech)

in 2004. They have been surprisingly long-lived and in most cases survived longer than expected. The power supply has normally been driven via solar cells, but this approach has set some limits. Naturally, the panels do not work at all during night time, and produce much reduced power in shadow. But more seriously, a covering of dust has resulted in permanently reduced efficiency.

So far, the only European attempt with a soft landing vehicle ended in failure. In 2003 Beagle 2 was supposed to bounce softly down on the surface, but something went wrong and all communication stopped just after entry into the atmosphere.

The latest NASA Mars rover, Curiosity, is equipped with a radioactive battery giving a considerably higher and more stable power supply than solar cells. In August 2012 it landed without problems, and it is currently exploring the planet. In 2015 it was still successfully on its way to Mount Sharp at the centre of its landing site, inside the Gale crater. So far it has confirmed that environmental conditions do not rule out earlier primitive life forms, but conclusive evidence for discovery of life remains to be found. Further light might be shed in 2016, when NASAs next Mars lander, InSight, is planned to arrive. In an international collaboration, it will drill into the Martian soil, equipped with seismometric and heat sensing detectors.

Several space probes have orbited the planet at length, investigating it with a variety of measuring devices. In particular, large numbers of spectacularly detailed images have been recorded. Specialised high resolution cameras are capable of showing details down to less than one meter in size. A notably successful mission in this category is the Mars Reconnaissance Orbiter which arrived in 2006 with an expected mission time of two years. In 2015 it is still operational and has discovered the destiny of the unfortunate Beagle 2 expedition. It now appears that the European vehicle successfully landed, but some of its panels did not unfurl correctly, blocking the communication system. The latest addition to Martian research is the Indian Mars Orbiter Mission, which arrived in orbit around the planet in September 2014.

A complete Martian atlas (e.g. GoogleMars) is now available. A glance at these internet-accessible images could well give the impression that Mars is documented at least as well as the Earth. As a consequence of all this new information the knowledge about Mars has deepened in a remarkable way. Our modern view of Mars now gives a rather complete picture, which nevertheless still has room for surprises—not least regarding the fundamental question about possible life.

The Martian Environment

Mars has an atmosphere, but unlike the Earth's it is very thin. It contains 95 % carbon dioxide, 3 % nitrogen and 1.6 % of the noble gas argon. In addition, there are very small quantities of oxygen and water vapour. The pressure is only half a percent of the Earth's, corresponding to an altitude of 35 km on Earth. The atmosphere also contains a lot of dust, which makes the Martian sky reddish as seen from the surface. The dust sometimes can form into giant dust storms (see Fig. 4.9). The planet was completely covered in dust when, in 1971, Mariner 9 became the first space probe to go into orbit around Mars. After about one month the view cleared and the investigations could start. Although the atmosphere is quite thin, it still shows meteorological activity in the form of weather systems. But weather effects on Martian geology are very small, which together with a generally low geological activity implies that erosion on Mars is also small, just a little greater than on the Moon. Consequently, what we see today on the Martian surface is preserved from a very early phase in the history of the planet. This distinguishes it from both Earth and Venus. Although thin, the atmosphere does exhibit surprises. For example, from the surface, the Phoenix lander could image snowfall, which dissolved before reaching the ground. A further interesting fact is that small amounts of methane gas have been found in the atmosphere. This gas is unstable, meaning that it breaks down and disappears with time. That it can be observed in the atmosphere indicates that it is replenished continuously in some way. There are a number of possible explanations, among them that micro organisms could be the cause.

The temperature on Mars is on the average about –55 °C, which is considerably lower than on Earth. Since Mars does not have a dense atmosphere or large oceans in order to keep and distribute the heat, like the Earth does, temperatures on Mars vary heavily, more than a hundred degrees during the length of a full day. Although temperatures up to and above 20 °C are possible at the equator during the summer season, more often a biting cold far below –100 °C dominates. Another important discovery which was made early on is that Mars, unlike Earth, has no magnetic field. We will return to the consequences of this condition.

When the shock after seeing the crater-filled landscape images from Mariner 4 died down, Mars turned out to have a quite varied and interesting topology. Clearly, there were thousands of craters, but another, more suggestive, discovery soon followed. Images showed numerous traces of flows of liquids, everything from dry river valleys to large areas looking like sea floors. Figure 4.8 shows some examples of this. Evidently, these are strong indications that Mars must have had a very different climate and appearance in the past. Interestingly, the surface is separated into a pronounced highland part in the South and a corresponding lowland part in the North. The topography is quite reminiscent of continents and

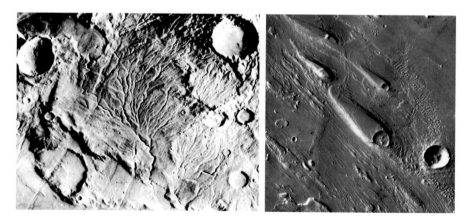

Fig. 4.8 *Left*: An early close-up of Mars. The craters makes the surface appear similar to the Moon, but obvious traces indicate previous existence of liquids. *Right*: Powerful streams have shaped the Martian terrain in prehistoric times (NASA NASA/JPL-Caltech/ASU)

ocean basins of Earth. The highlands are clearly older than the lowlands since the number of craters is larger. In the lowlands the surface is notably smoother. The explanation may be giant impacts which created enormous basins of floating lava which subsequently solidified. These were probably caused during a period which goes under the name Late Heavy Bombardment about 4 billion years ago. The solar system had formed about 600 million years previous to that event. Similar basins are also found on the Moon.

The Martian "landscape" that was observed from the Earth over the course of several hundred years was quickly demystified. The light red land areas simply consist of fine dust while the darker "continents" are areas where the wind has swept the dust clean, making the underlying rock more visible. The changes in the pattern observed from Earth clearly were not, caused by vegetation, as the optimists believed, but from winds randomly moving around the dust layers.

A number of giant volcanoes bear witness to a high degree of geological activity during a very early epoch. Olympus Mons on Mars is largest known volcano in the solar system, with a diameter of 600 km and a height of 22 km. That the volcanoes are so much larger than on Earth is explained by the fact that plate tectonics (which causes the continental drift on Earth) seems to have all but disappeared on Mars. On the other hand, very strong tensions must once have been present in the surface layers on the planet. One proof of that is one of the mightiest landmarks on Mars, the gigantic ravine Valles Marineris. It is over 4000 km long and at some places 10 km deep.

In spite of its currently low geological activity, Mars is not a static planet, but changes in various ways. We have already mentioned the dust storms which affect the appearance of the landscape. More local "dust devils" have been observed several times. These are strong whirlwinds that are also present on Earth. Even more spectacular are the enormous landslides which happen occasionally. In 2008 one Mars probe captured unique pictures of an avalanche in progress (see Fig. 4.9).

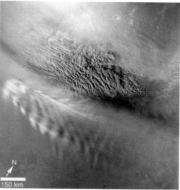

Fig. 4.9 *Left*: An ongoing landslide on Mars, as seen from above. At the *left* edge ice coverage is seen. The Martian soil has just fallen a precipice of about 700 m height. *Right*: A gigantic dust storm, caught in an image taken by an orbiting spacecraft (NASA NASA/JPL-Caltech/MSSS)

Is There Any Water?

The big question is whether there is or has been water on Mars. This is really the essence of—and target for—most of the Mars research over the most recent years. NASA uses the motto "Follow the water!" for its Mars exploration, with the goal that in the end the strategy will lead to the answer about life on the planet. And scientists nowadays agree that there must have been enormous quantities of liquid water on Mars; there is no other feasible explanation. In addition, large amounts of water still exist, although not on the surface in liquid form. Scientists have now concluded that the polar caps of Mars are, to a large extent, made of water ice with a smaller contribution of dry ice (solid carbon dioxide). The dry ice lies on top in a rather thin layer. The coverage of this layer is strongly influenced by seasonal changes. During Martian winter the temperature falls so much that a portion of the carbon dioxide in the atmosphere is deposited as dry ice. When spring arrives, the dry ice sublimates, i.e. transforms directly into gaseous form again. This annual change in the appearance of the polar caps can be observed from Earth, even in small telescopes. The total water ice quantity which is bound to the polar caps is estimated to about

3 million cubic kilometres. This can be compared to the ice on Greenland, which contains about 2.8 million cubic kilometres.

It is also probable that large amounts of water reside beneath the surface, in the form of permafrost or even in floating form. There are several indications that water that has recently broken out of the surface; distinct traces can be found along some mountain ridges. Some scientists claim that Mars has a groundwater level and that emerging water may have contributed to the mineral deposits found in some places, not least inside craters.

The Fourth Experiment

There are other observations that point towards the direction of primitive Martian life. The first Mars landers, Viking 1 and Viking 2, had specialised experiments on board in order to try to identify primitive life. Four different experiments were carried out at the two landing sites. Three of them quickly gave negative results. The fourth experiment sensationally enough gave a positive result. Some Martian soil was soaked in a modified nourishment solution where normal carbon atoms (^{12}C) were exchanged for radioactive carbon (^{14}C). The idea was that if some sort of metabolism took place, the gas released from the heated specimen would contain the radioactive carbon atoms. It turned out that radioactive gas could be detected immediately. The result was identical for both sites. The excitement was great but somewhat reduced when the experiments were repeated a week later and then gave a different result. So the three first experiments were negative. What had happened in the fourth experiment? Well, expert opinion differ, in a way often typical for scientists, and a definite answer is not available. The majority, however, assert that explanations other than the existence life are more probable. The detected effect may have been due to non-biological chemical reactions.

The Mysterious Meteorite

One day, about 13,000 years ago, a very strong light illuminated the sky above the Antarctica. Aborigines in Southern Australia may

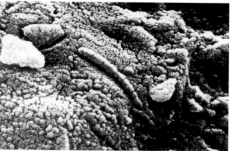

FIG. 4.10 *Left*: A meteorite with its origin from Mars. *Right*: The controversial picture of a possible organism found in the interior of the meteorite (NASA/LSC/Stanford University NASA)

have been able to see it. It was a large meteorite that fell to Earth, completing a 16 million year journey through space. Remarkably enough, the two kilo meteorite (see Fig. 4.10, left) stayed on the ice surface until 1984 when a group of American geologists finally encountered it. The meteorite was given the designation ALH8400. It remained unnoticed in storage for several years before it was inspected in closer detail. Analysts determined that small, embedded gas bubbles contained exactly the same gas constituents that had been measured on site in the Mars atmosphere by the Viking landers. It was clear that the meteorite originated on Mars. This was an exciting discovery, although not that uncommon, since meteorites from Mars were known before. But it would turn out that this meteorite was unique. In a report published in 1996, a group of NASA scientists claimed that the stone contained clear evidence of previously existing Martian life. This caused a great sensation, and US president Clinton made a somewhat premature statement in which he explained that the discovery potentially was one of the greatest ever in the history of mankind. The claim of the scientists initiated an intense debate regarding the validity of the evidence that had been exhibited. There was consensus about a few basic facts: the stone had undergone a rather adventurous existence. It was very old and had been formed on Mars about 4 billion years ago, simultaneously with the general formation of the planet. Almost a billion years later, it was subjected to strong forces from a large meteorite impact on Mars. Thereafter it was exposed to the Martian weather and climate conditions of the time, experiencing the wet period of Mars. Much later, about 16 million years ago, it was subjected to another meteorite hit which hurled it into space. It remained in space for a few million years. During this time it was affected by solar wind and cosmic radiation before chance put it in the path of Earth, where it found a new resting place. Finally, during the last 13,000 years, Earth's environment, albeit under Artic conditions, affected it in various ways.

The group behind the sensational report claimed to have seen a number of features on and inside the meteorite which, they suggested, were best explained by life forms on Mars. They felt that life could have been there when Mars was a young planet. Without going into too much intricate detail in chemistry and microbiology, their discoveries can be summarised in the following ways. They found:

- calcic globules, reminiscent of similar objects which on Earth are created in water by biological processes
- magnetite crystals, reminiscent of those which on Earth are created by certain marine bacteria
- amino acidic substances that could be a side product from disintegration of biological molecules

However, the observations that created the most excitement in the world media were images taken during the electron-microscopic investigation at very high magnification. They showed something sensational that could be interpreted as a kind of bacterium (see Fig. 4.10, right). A fierce debate erupted with different interpretations of the results. In addition to pure scientific arguments, prestige played a significant role. A Nobel prize would be awaiting the first scientist proving the existence of alien life. So, alternate explanations were offered at every point in the discussion. The problem with the look-alike bacteria was that they were smaller than their supposed terrestrial counterparts, only 20 nm wide, or 20 millionths of a millimetre. That size bacteria do not exist on Earth, even if some scientists disagree with that view.

As is so often the case in scientific controversies, the experts share different views. The majority has so far dismissed the possibility that ALH8400 proves that life once existed on Mars. But science builds on observations that in turn give new knowledge.

There are currently studies in progress of other Martian meteorites; at least twenty are known to exist. Perhaps the opinion will turn again. Naturally, to make a manned expedition to Mars and bring back a few kilos of selected Martian soil would most probably clarify the matter.

The Tragic History of Mars

It is obvious that not only Venus but also Mars must have gone through dramatic changes during the history of our planetary system. Circumstances on Mars were radically different when the planet was young. It is difficult to avoid the thought that provisions for the emergence of life may have been very good in the past. We have seen that Venus encountered a catastrophe in the form of an accelerating greenhouse effect, but what went wrong on Mars? That we today can see both craters and abundant evidence for flowing water gives us clear clues to the story. First and foremost, the current appearance of the Martian surface must be very old. Large meteor and asteroid bombardment happened early, at least more than 3 billion years ago. The traces of this era are seen all over the solar system, with Mercury and our Moon as prime examples. The fact that Mars once harboured large quantities of liquid water must mean that it once had an atmosphere with sufficient pressure. This atmosphere probably consisted of carbon dioxide, and would have created a greenhouse effect that very well would have given the past Mars a much more comfortable climate than now (see Fig. 4.11). The Mars of today is more or less lacking a magnetic field, but studies imply that there was one in the beginning. Traces of lingering magnetism have been found in minerals. There are even indications of changed polarity in a way similar to that seen in the behaviour of Earth's magnetic field. At one time, a working dynamo seems to have been operating on Mars, generating a protective magnetic field. It has not been confirmed whether plate tectonics ever existed in order to move around the Martian surface and create geological formations in the crust. But the enormous volcanoes suggest otherwise. They have been able to grow precisely because the Martian crust beneath them has not moved. On Earth the opposite is seen, as volcanoes are often formed in groups, like the Hawaiian islands.

Fig. 4.11 A possible Mars about 4 billion years ago—a water rich planet?
(Christian Darkin/Science Photo Library/IBL)

It has been suggested that the magnetic field may have disappeared very early due to huge impacts on the surface. These could have heated the surface layers to such an extent that the temperature gradient necessary to drive the convection between inner and outer layers disappeared.

The existence of a magnetic field is very important for a life-bearing planet. As we have discussed, it shields the planet and its atmosphere from electrically charged particles coming from the Sun, and from the cosmic background radiation. Because of its ionising effect such radiation (similar to radioactivity) is dangerous for living organisms, and it has been estimated that on Mars today a hypothetical DNA-based life would have to be at least seven metres below the Martian ground in order to survive.

How did the atmosphere of Mars disappear? Cosmic radiation plays a role. The high energies in the radiation disassociate the molecules of the atmosphere, and facilitates its disappearance into space. Together with the lower gravitation of Mars and, hence, lower escape velocity (5 km/s), this probably explains how the atmosphere successively was thinned out to its present level. But the mystery is not completely solved. So, in September 2014 NASA's latest Mars probe, the Mars Atmosphere and Volatile Evolution (MAVEN) arrived in orbit around Mars. Its mission is to explore the planet's upper atmosphere, ionosphere and interactions with the sun and solar wind. Early results from December 2014 seem to corroborate the current scenario reasonably well.

Today's Mars clearly is not a place well suited for higher forms of life. In spite of this, dubious speculations of an advanced Martian civilisation—alive or dead—still appear. In Fig. 4.12 we see a classical example, the infamous Martian face. The picture to the left was taken by the Viking orbiter in 1976 and is a small detail in a survey image. Heavily magnified and somewhat enhanced, a rock formation appears reminiscent of a human face. A cult was created around this; some people preferred to interpret

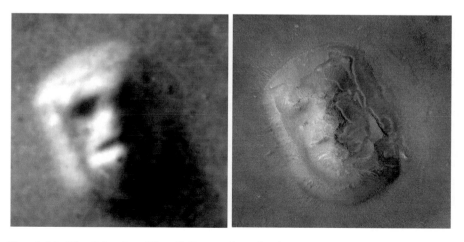

Fig. 4.12 The Martian "face". Even after the first Mars expeditions, speculations about messages to Earth from Martians still flourished. The rock formation to the left was by some interpreted as a human face, intended as a message. To the *right* the same formation is seen from a similar angle but with much higher image resolution (NASA)

this as a kind of message to the civilisation on Earth, something that Percival Lowell in the nineteenth century surely would have appreciated. But now we are in the twenty-first century. The picture to the right was taken in 2001 by Mars Global Surveyor, with a much higher resolution. With this evidence, even the hardcore Martian fan must yield.

The question as to whether life in primitive form exists or has existed on Mars is still wide open. A definitive answer should be available in the near future—at the very latest when man himself will have the opportunity to visit the planet, probably sometime around 2030.

Life in Unexpected Places?

We note that the three planets that actually are inside the habitable zone of our solar system display three very different constitutions. Both Venus and Mars have, each in their own way, become victims of a catastrophic development. So should we draw the somewhat pessimistic conclusion that the creation of sufficiently stable environments for the emergence of life is a delicate act of balance and therefore a very rare event? No, such an assumption would be premature. Additionally, there is now hope that new information from planets in other star systems eventually will throw new light on this issue. But there are actually other discoveries in our solar system that expand our perspectives even more.

The normal definition of a habitable zone is based upon the supposition that the nuclear energy developed inside stars is the fundamental energy source necessary for life to develop on any planet. But there are other energy sources in the solar system, albeit weaker but possibly almost equally as important. First of all, some of the heat generated during the original formation of the planets remains. During this formative time, giant collisions were very frequent.

Secondly, there are large amounts of energy coming from radioactive isotopes which gradually decay and dissipate heat. In the case of the Earth, it is believed that about 20 % of the inner heat of the Earth comes from elements like uranium and thorium.

Thirdly, we have the gravitational energy which in suitable circumstances can be converted into mechanical energy, and from there to heat. The most well-known example is probably the tides of Earth's oceans. Tides are caused by the gravitational attraction from the Moon and to a lesser extent from the Sun. It is the difference between the lunar gravitation at the centre and the surface of the Earth that creates the tidal effects. We are talking about considerable amounts of energy: roughly 3000 Gigawatts are created (more than 1000 times the total output of all the nuclear power plants on Earth). This, has the side effect of decreasing the rotational speed of the Earth about a second every 100,000 years. The distance to the Moon increases by about 4 cm every year.

The tidal effect is not unique to the Earth-Moon system. But it still came as a surprise when the first close-up pictures of other moons began to appear on screens on Earth. In a series of brilliant space explorations in the 1970s and 1980s, the USA sent the Pioneer and Voyager spacecraft towards the outer regions of our solar system. The spacecraft studied the giant planets Jupiter and Saturn, and in a "grand tour" Voyager 2 was sent even further towards Uranus and Neptune. The results were in many ways amazing. But most spectacular of all were the close-ups taken of the moons of Jupiter—they were sensational.

Jupiter has many moons, but four of them are paramount. They are called the Galilean moons, as Galileo could see them in the seventeenth century. Actually, today everyone can see them: all that is needed is a pair of powerful binoculars. But seeing something from Earth as more than a small bright dot is very difficult. Yet, in reality the moons are quite large. The innermost two, Io and Europa, are the same size as our Moon, while the two outer ones, Ganymede and Callisto, are larger, more like Mercury in size. Their respective distances to Jupiter are between a half and two million kilometres. This is rather close, especially considering the enormous gravitational pull from Jupiter: its mass is more than 300 times that of the Earth.

One could expect large tidal effects on the moons, but the images received from Voyager were, as mentioned, sensational. Volcanoes covered the innermost moon, Io. The Voyagers even imaged an ongoing volcanic eruption, which certainly was as far as you could get from the conventional picture of an ice cold moon very distant from the heat of the Sun. Io was continually moulded by the enormous tidal effects, as Jupiter's gravitational field heated the entire moon and caused the continuous volcanic eruptions. With this it became clear that conditions for life are not necessarily connected to a suitable distance to a sun. However, in the specific case of Io conditions were more hellish than paradisiacal.

The moon Europa, a little more distant from Jupiter, was interesting in another way. The best images of Europa came in 1995 from the spacecraft Galileo, which then entered into orbit around Jupiter. The spacecraft performed a series of close passages of Europa. At its closest it was a mere 2000 km from its surface, revealing details down to a size of only 20 m! Figure 4.13 shows that the moon appears covered in ice and has some grooves in the

FIG. 4.13 Top: Jupiter's ice-covered moon, Europa. Bottom: The close-up shows that the surface is criss-crossed by ice fractures (NASA/JPL/DLR/ University of Arizona)

surface. In the close-up we see a chaotic terrain completely criss-crossed with a lot of apparently random structures.

At close range a new world emerged, with no equivalent elsewhere in the planetary system; an ice-covered and cold world, but nevertheless geologically very active due to the tidal effects from its proximity to Jupiter. These effects cause continuous rifts in the ice crust which subsequently freezes over again, a process that repeats itself thousands of times, creating strange and complicated patterns in the terrain. According to some researchers, the surface of Europa may be no older than 100 million years. The moon's youth is also indicated by the fact that there are rather few craters on Europa, they are relatively small and seem to be newly formed.

The surface climate of Europa is not hospitable for life. The temperature at the surface is about –160 °C, and it is continually subjected to intensive radiation from charged particles coming from the radiation belts of Jupiter. An unprotected human would quickly receive a lethal radiation dose and not survive for long. But beneath the ice crust, conditions are different. The inner parts are warmer and well protected from the radiation. Indications are that water exists in liquid form, probably within a 100 km deep ocean, on which the entire ice crust is floating. Although Europa has a bound rotation with respect to Jupiter, there is evidence that the actual surface layer is movable, independent of the central core, which in itself is evidence for the existence of a liquid beneath the icy surface. A weak magnetic field has been detected and scientists believe that this is generated through motions in a saline water layer.

In Europa, we have a new candidate for alien life, far from the habitable zone. Mars may have had a hospitable environment some billion of years ago; beneath the ice of Europa such an environment seems to exist right now. But no concrete signs of life have been discovered. If it exists, it would then be hidden far below the ice, a difficult place for observations. Incidentally, on Earth there is a similar environment in the form of the Vostok Lake, which has been completely encapsulated under the Antarctic ice for thousands of years. The exploration of Europa is still in very early stages, and there is a great interest in further studies.

The European Space Agency (ESA) has recently decided to send a spacecraft to closer investigate the moons of Jupiter, and NASA is currently studying a similar concept. The aim is set for Europa, Ganymede and Callisto. Ganymede and Callisto may also have liquid water below their hard ice crusts. The ESA expedition has a planned launch for 2022 and arrival in the Jupiter system in 2030. Perhaps the probe will be able to clarify if there is life on Europa. Otherwise we will have to wait a little longer. The next step will probably be to send a probe to land on Europa and drill a hole into the ice, on site, in order explore the extraterrestrial ocean.

Exotic Environments in the Planetary System

Saturn has a large moon called Titan. In the entire solar system, it is second in size only to the Jupiter moon Ganymede. Titan is the only moon in the solar system with an atmosphere, quite dense and impregnable for light. In this aspect it is similar to Venus. To see anything you need either radar studies from outside or to actually land on the surface. In 1997 the spacecraft Cassini-Huygens was sent on a long mission targeting Saturn. Basically a NASA project, ESA made the important contribution of constructing a Titan lander, Huygens, named after the discoverer of Titan in seventeenth century. The spacecraft followed a complicated orbit to reach its goal. First it was sent inside our solar system. After two passages of Venus it accelerated outwards, using gravity assist. Passages of the Earth and of Jupiter further increased the speed before final course was set for Saturn. This is an economical way to travel in the solar system, although it takes considerably longer time. However, also during flight things can be achieved. The Jupiter rendezvous resulted in the most detailed images so far taken of the planet. On the way out to Saturn an exciting experiment was performed in order to test Einstein's theory of general relativity, which predicts that space-time around the Sun is curved by the gravitation of the Sun. The experiment was carried out when the Cassini-Huygens craft had reached a position behind the Sun when viewed from the Earth. This means that its radio waves were travelling a longer path in curved space-time than would otherwise have been the case. This curvature could be measured with a very high accuracy. The result completely corroborated Einstein's theory. Eight years later, the craft finally reached its target and entered orbit around the ring-adorned planet.

The most dramatic and historic part of the journey was the soft-landing on the Saturn moon Titan. On Jan 14, 2005, the Huygens probe detached from its host vehicle and glided into the Titan atmosphere. Almost everything went according to plan and a new world was revealed before our very eyes. In Fig. 4.14 we see a picture of Titan taken by Cassini, as seen through the rings of Saturn. Figure 4.15 shows the landscape photographed by Huygens after the landing. In addition to the results from the successful

FIG. 4.14 Titan imaged by the Cassini probe, with the rings of Saturn and the small moon Epimetheos in the foreground (NASA/JPL/Space Science Institute)

Fɪɢ. **4.15** *Left*: The Titan landscape as seen from the Huygens lander. *Right*: The surface landscape as reconstructed from radar measurements. The dark areas correspond to lakes of liquid hydrocarbons (NASA/JPL/Caltech/USGS)

landing, Huygens also studied Titan with several instruments, among them radar from which survey images of the surface could be acquired.

The scenery emerging was not totally unlike some places on Earth, with the notable difference of a temperature about 200° lower than on Earth. The atmosphere is actually a little denser than the Earth's, and consists primarily of nitrogen. It is opaque for ordinary light, because ultraviolet radiation from the Sun creates various hydrocarbons in the upper layers of the atmosphere which appear similar to a dense smog.

Most exciting of all, Titan seems to have a landscape with mountains, plains and lakes. There is also strong geological and meteorological activity. But the actual chemistry is quite different. The ground formations are not made of minerals but probably from water ice, the lakes and rivers do not contain liquid water like on Earth, but instead are made of liquid hydrocarbons like ethane and methane. It is possible that beneath the rigid surface of Titan there exists an ocean of a liquid. In such a case it would probably be a mixture of water and ammonia.

Can there be any life in such an extremely different environment? There are scientists who believe there is. The discovery of a world like Titan's has reinforced the debate over whether life can manifest in other forms than our own. It is a fact that many complicated molecules have been found in the atmosphere of Titan, which is chemically very active. In some respects the conditions on Titan are reminiscent of those that existed when life began on Earth, with the important exception of liquid water. But there are speculations about whether life may have found other solutions. Perhaps the rivers of ethane harbour life that can utilise hydrogen instead of oxygen and react with acetylene instead of glucose and as waste product produce methane. A hint in this direction was provided to scientists when they discovered that the content of hydrogen was considerably higher in the upper parts of the atmosphere than in the lower parts. Perhaps the hydrogen at the surface was consumed by exotic micro-organisms?

In a different scenario, Titan might become a living world in the far future. When the Sun reaches its final development phases in about 4 billion years time, it will swell to a giant red star. The Earth will then perish but the climate on Titan will become quite comfortable during a few hundred million years. Perhaps it is then that life will emerge and flourish on Titan, but we will probably not be around to observe it.

In conclusion, only a minority of optimists believe in life on Titan. Mars is undoubtedly the planet which has to be considered as the main candidate for extraterrestrial life in our solar system. But the best we can probably hope for is to find traces there from earlier life. Even so, this would of course be a tremendous discovery. But on the other hand, our perspectives of what life can be like are expanding all the time, and surprises can be awaiting us even in our own solar system.

5. Exoplanets—From Speculation to Reality

A few times in the history of man, astronomers have made discoveries influencing people's conception of the universe and of themselves. The Greeks of antiquity were known mostly for their theoretical and philosophical reasoning, but some also carried out experiments. In about 240 AD, Eratosthenes not only deduced that the Earth was spherical, but also determined its circumference with impressive accuracy. He used simple but ingenious observations of the shadows that the Sun casts at different latitudes at the same time. In 1543 Copernicus finally published his suggestion that the Earth actually was not at the centre of the universe. This was soon proven by Johannes Kepler with the help of observations made by Tycho Brahe; the Earth was clearly in an elliptical orbit around the Sun. The realisation that stars are like suns but at very large distances also belongs to these revolutionary insigths. Another upheaval came with the discovery that many diffuse light patches turned out to be galaxies containing billions of stars. Yet another was the conclusion that the universe began with a Big Bang some 14 billion years ago. Now, mankind is on the threshold of another paradigm shift in its conception of the universe. Are there other planets like our Earth, and if so, is there life on such planets, even intelligent life?

On October 6, 1995, the world history changed yet again. For ordinary people this date passed inconspicuously, but it was a great day for the Swiss astrophysicists Michel Mayor and Didier Queloz at the Geneva observatory. At an astronomy conference in Florence, they presented their observations of the star 51 Pegasi, a star quite similar to our Sun. They had found evidence that there was an object affecting the star. Their conclusive evidence was shown in a figure that revealed very small but periodic movements of the star. The reason for these movements could be none other than the influence from an orbiting planet. With this the speculations

© Springer International Publishing Switzerland 2016
P. Linde, *The Hunt for Alien Life*, Astronomers' Universe,
DOI 10.1007/978-3-319-24118-0_5

about planets around other stars had gone from science fiction to science fact. And a new word was coined, exoplanet—an abbreviation of the term extrasolar planet, i.e. a planet belonging to a star other than our Sun. What had made it possible to establish, at last, the existence of a planet around another star?

The Fixed Stars Are Not So Fixed

The idea that a planet orbiting another star could reveal its existence by its effect on the star is not new. That the "fixed stars" actually moved with respect to each other had been determined in the nineteenth century. Such movements are explained by stars having their real own motions, quite often in the order of a few tens of kilometres per second relative to our solar system. The corresponding apparent motion in our sky depends on the distance to the star and on the star's direction of motion. But it is always very small.

In 1916, the American astronomer Edward Emerson Barnard measured the change in position of the star that still has the largest known proper motion of all observed stars. The proper motion is expressed in angular motion per unit time, and we are talking about small angles. Barnard's star is moving 10.3 seconds of arc every year. After about 180 years this corresponds to the angle subtended by the Moon in our sky. However, most of the stars we can see by naked eye have much smaller apparent motions than that. Launched in the 1990s, the Hipparcos satellite made dedicated and highly accurate determinations of stellar positions. The project became a major scientific success, resulting in knowledge about exact positions and motions for all stars within the nearest 400 light years. A resulting example is seen in Fig. 5.1 which shows how the stars of the Big Dipper move over the course of 100,000 years.

How Do You Discover an Exoplanet?

Astronomers assumed that one way to detect planets would be to look for possible deviations in the motions of the stars. The standard conception of a star-planet system is that the planet moves around the star in a more or less elliptical orbit. But in fact a planet also influences its host star through its gravitation, like a moon does

FIG. 5.1 The constellations on the night sky change slowly due to the proper motion of the stars. To the *left* is shown the Plough 50,000 years ago, in the *middle* its current appearance and to the *right* what it will look like in a further 50,000 years (Peter Linde)

its host planet. One example of this is the Earth-Moon system. The two move around a common centre, which actually lies inside the Earth, not so far from the centre of the Earth. A corresponding gravitational pull from a planet creates a waggling effect on its host star. The larger the planet, the larger the effect becomes—but it is, in any circumstance, quite small. You can get an appreciation of what can be expected from studying these effects at close range, in our own solar system.

Figure 5.2 shows the motions that our Sun makes under the influences of its planets. Every dot in the figure shows the position of the common centre of gravity for the entire solar system during the years 1945–1995. The Sun itself moves around this point in space. A closer study of the figure shows that the Sun sometimes moves around a position outside its own radius with a period of about twelve years. This corresponds to the orbital period of Jupiter, the largest planet, which happens to be twelve years. Smaller effects from other planets are added to this motion, mainly from Saturn. Seen from a distance of for example twenty light years, this motion creates a maximum angular sideways difference of three milliarcseconds. That is an incredibly small angle, comparable to the diameter of golf ball as observed at a distance of 3000 km!

From this it is also possible to estimate the maximum motion of the Sun due to gravitational effects from the planets. That comes out at a little less than 10 m/s, which corresponds to a comfortable cycling speed. In astronomical contexts this would normally be considered as vanishingly small. Naturally, the effects from a much smaller planet, such as the Earth, would be even less, about 1000 times smaller!

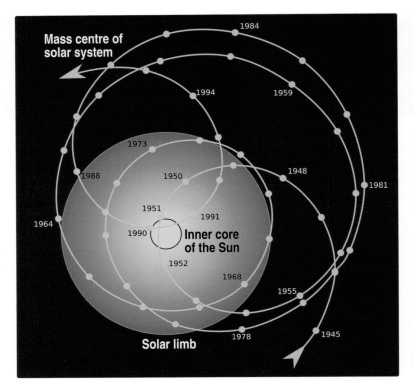

FIG. 5.2 The planets around the Sun, especially Jupiter and Saturn, affect the position of the Sun with their orbital motions. The figure shows this effect during a 50 year period. Every dot corresponds to the common gravitational centre of the solar system a certain year. The Sun evidently wiggles in its position, sometimes even around a point outside the solar surface (Peter Linde)

Are such things really measurable? The answer is yes; with the technology of today it is possible to attack even such difficult challenges. But when we are dealing with extreme measurements of this kind, it turns out that it is actually easier to measure a velocity than to measure a position. Let us take a look at the technology available.

Technology Is the Key

The continuous development and improvement of instruments and telescopes have always played a decisive role in our increased knowledge about our expanding universe. Tycho Brahe and Galileo were groundbreaking pioneers, and from then on revolutionary discoveries have been made almost every time a new and larger observatory has become operational. This trend continues with ever-newer technological advancements. Advanced mechanics and optics, combined with the latest progress in computer science, have led to new revolutionary designs. To this should be added the development on the detector side, where the Nobel prize-awarded invention of the CCD camera has reached a level where it is now possible to detect almost 100 % of all incoming photons. The previous photographic technique was capable of only registering a couple of percent of the light. With these modern detectors, the remaining option to further improvements relies on an increase in the size of the telescope mirror. This is exactly what is currently under way.

Hence, astronomy is a very technology-dependent science. With the notable exceptions of the space probe visits to the Moon and parts of the solar system, astronomers need to rely completely on the passive capture of the information that reaches Earth in the form of various types of light. In contrast to many other fields of natural science, astronomers cannot adapt reality in the form of suitable arranged experiments. Instead they need to make sure that the light that has been on its way to Earth for thousands, millions or even billions of years is being received and analysed in the best possible way.

First of all, you need a large and powerful telescope. Space-based telescopes have a great advantage: with no atmosphere to look through, images are not degraded by turbulence in the air. Also, space telescopes have access to the entire spectrum of light, not just the part that filters through down to the ground. But there are also considerable disadvantages. Space-based instruments are very expensive to manufacture and to launch. There are limited possibilities, if any, to repair or update them once they are in space. The famous Hubble Space Telescope (HST) is an exception. It was designed from the start to be revisited and updated via a space shuttle mission. This was, by the way, well advised since a revisit was needed immediately, because the mirror of the telescope turned out to be ground and polished to the wrong shape. The optical error could fortunately be well characterised from the ground, and three years after the initial launch a group of astronauts succeeded in installing a specially made optical correction system.

A catastrophic failure was then changed into a great success. Since the repair mission, Hubble has taken the sharpest images ever. But normally satellites must function perfectly from the beginning and continue to do so for years. In the future this will become even more important since locating satellites at distances far from the Earth will become more commonplace.

However, ground-based observatories continue to compete. The factors that are disadvantages for space telescopes are advantages for ground-based ones. Above all, building large constructions is much easier if they can stay on the ground—and size matters. Therefore, a new and revolutionary type of telescope is being designed. We are referring to the extremely large telescopes, to which we will come back in a while.

Can You Measure the Speed of a Snail—10 Billion Kilometres Away?

A determined but ordinary garden snail moves at a speed of about 1–2 mm/s. Let us engage in a thought experiment: we will transfer the snail about 10 billion kilometres away (about 60 times the distance between Earth and Sun) and let it crawl there perpendicular to the line of sight. Could this in any way be measured? At this speed and at this distance our snail would move as slow (measured as angular speed) as a star does at a distance of 10–30 light years while being affected by a reasonably large planet. Of course the snail can never be observed, but remarkably enough such a small effect on a star can be measured.

Fortunately, nature has provided us with a shortcut. What we can measure with immense precision is not the perpendicular shift but rather the motion along the line of sight. Figure 5.3 illustrates a star that is affected by a planet in such a way that the star is waggling a little. This is really just a simplified version of the situation shown for our own solar system in Fig. 5.2. To the left is marked the small angle that the star is moving. We can measure such an angle, but it is exceptionally difficult to do. Nevertheless, the new Gaia satellite is equipped to do this (more about this in Chap. 7). The right side of Fig. 5.3 illustrates our shortcut, namely a measurement of the star's motion along the line of sight. This measurement is based on something known as the Doppler effect.

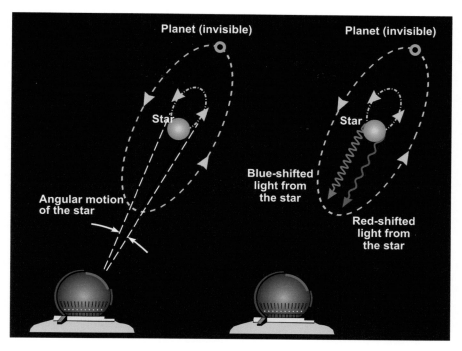

FIG. 5.3 Two methods to discover exoplanets. To the *left* is illustrated measurements using astrometry to detect stellar sideway movements, to the *right* using the Doppler effect to detect motions along the line of sight (Peter Linde/PHI)

In 1842 the Austrian physicist Christian Doppler published a scientific paper in which he described the rather mundane physical phenomenon which would eventually be named after him. The phenomenon applies to all kinds of wave motions, sound as well as light. Most people recognise this from the audio world. An ambulance with an active siren passing a stationary person on a road may serve as an example (see Fig. 5.4). As long as the ambulance is moving towards the person its siren has a certain pitch, which is the frequency of the sound (number of sound waves per unit of time). This frequency is higher than the frequency generated in the siren, because the sound waves of the siren are compressed in the direction of motion. The degree of compression depends on the speed of the car. At the actual moment when it passes the person, the person notices a characteristic lowering of the pitch (even if the actual volume, the sound level, increases). After passing, the sound waves are instead drawn out relative to the observer, which results in a lower pitch and frequency.

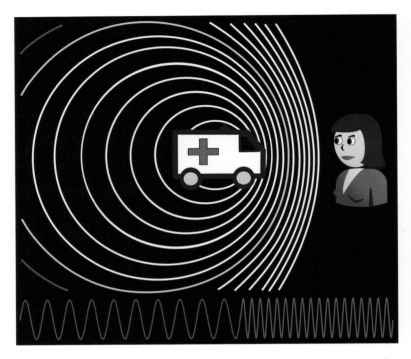

FIG. 5.4 An ambulance on its way towards an observer. The sound waves from the siren are compressed to a higher frequency. When it has passed the waves get extended (Peter Linde)

Exactly the same thing happens to the light coming from a star. If the star is moving away from us, its light will have a decreased frequency, if it is moving toward us the frequency increases. Since the frequency and colour of the light are closely interconnected, this corresponds to a red and blue shift, respectively, of the wavelength of the light. Interestingly, such measurements can be made with incredible accuracy, better than one part in 100 million. However, to do this requires access to a large and specially equipped telescope.

Measuring the Radial Velocity of a Star

In practice a star moves in all possible directions, but as can be deduced from Figs. 5.3 and 5.4, the Doppler effect can only be utilised for that part of the motion that lies in the line of sight, either

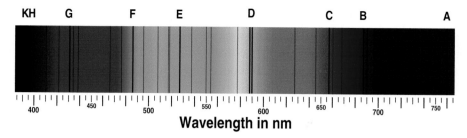

FIG. 5.5 A stellar spectrum for a solar type star. The *vertical lines* are so called absorption lines. Some of the lines, marked with *letters*, were catalogued already in 1814 by Fraunhofer. The positions of the absorption lines are displaced by the Doppler effect (Peter Linde)

forwards or backwards. We call this the radial velocity of the star. But how is the measurement made in practice? As is well known, ordinary light consists of a mix of several basic colours. In nature we see one example in the rainbow, an effect that is created in case of rainfalls when the light of the Sun strikes water droplets in the air at a suitable angle. Since light of different wavelengths (colours) are refracted differently in the droplets and subsequently reflected in different directions, it becomes possible by naked eye to see the spectrum of the light in the form of a rainbow.

With the assistance of specialised instruments, so called spectrographs, astrophysicists can disperse the light of a star in a similar way. Researchers connect a spectrograph to a telescope, which catches and amplifies the light. However, to measure a weak colour shift depending on the Doppler effect would still not have been possible if nature did not offer yet another shortcut. If you study the spectrum from the Sun in more detail it turns out that it is not entirely continuous. In Fig. 5.5 we see a (simplified) spectrum from a star of the same type as our Sun. In addition to the rainbow-like colours we also see black lines, known as absorption lines. Some wavelengths (colours) are simply missing. Such lines are caused by the existence of different basic elements in the outer layers of the star. The gradual understanding and interpretation of such lines is one of the major triumphs of modern physics and has given us deepened and detailed information about the chemical composition of stars (see also Box 5.1).

The important thing in our case is that the positions of the lines are also affected by the Doppler effect arising from the motion of the star along the line of sight. They serve as references to measure the wavelength shift. If the star moves towards us all lines are equally shifted towards the blue and they are shifted towards the red if it moves away. Inside the spectrograph, the stellar spectrum is compared to a laboratory spectrum from a gaseous substance in rest. By simultaneously comparing a very large number of spectral lines it is possible to achieve a very high precision of the determination of the line shift and thus of star's radial velocity. An important requirement is that you have a lot of light from the star, i.e. it should be a rather bright star and you should use a large telescope to capture the light.

In this way it has become possible to measure those extremely small movements of a star that is gravitationally affected by its planets.

In Fig. 5.6 we see the HARPS spectrograph, one of the most accurate instruments in the world for measuring radial velocities. HARPS is an acronym for High Accuracy Radial velocity Planet Searcher. With HARPS it is possible to achieve radial velocity measurements down to 1 m/s, which corresponds to slow walking speed. We mentioned above that the motions we need to measure

FIG. 5.6 The HARPS spectrograph, which has been used in a large number of exoplanet discoveries (ESO)

Box 5.1: What Is a Spectrum?

All light is generated in the quantum-mechanical world of the atoms. Electrons orbiting atomic nuclei may jump to new orbits with other energy levels. A jump to an orbit with lower energy causes a photon with a specific wavelength (colour) to be emitted. Conversely, an electron can absorb an incoming photon and jump to an orbit with higher energy. Such excitation may also happen in other ways, in particular via collisions between the particles contained in the plasma that forms a star.

In the hot interior of a star, all possible wavelengths of light are generated in this way. This creates a continuous spectrum displaying all colours. But before the light leaves the star it passes through its outer layers. These are cooler (although still several thousands degrees hot), and they absorb certain wavelengths, unique and well-defined for each basic element. As a result, small gaps called absorption lines, consisting of missing wavelengths, appear in the continuous spectrum. See Fig. 5.7 for an illustration.

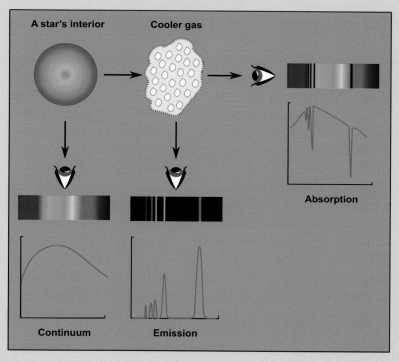

FIG. 5.7 The principle behind the creation process of different types of spectra (Eva Dagnegård)

Through the study of stellar spectra, astronomers gain detailed information on the composition of a star, its density, temperature and magnetic field. With the help of absorption lines, the Doppler effect can be used to study a variety of motions associated with the star. Examples of these motions include the star's motion in space, its rotation and its seismic vibrations.

are at a level of about a few meters per second. We should, however, be aware that looking for such weak signals is not an easy thing considering that there are a variety of motions present that influence the data and complicate the analysis, not least from the star itself. Nevertheless, HARPS coupled to the large ESO 3.6 m telescope in Chile has revolutionised the search for exoplanets (Fig. 5.7).

Stellar Eclipses: The Transit Method

Another way of discovering an exoplanet is if it passes in front of its host star, a transit as seen from the Earth. The light of the star diminishes during a short time, as a stellar eclipse takes place. This clearly requires that we observe such a system sideways, something that obviously will not be very common. But it does happen. In fact, we see this phenomenon happen in our own solar system. The planets that lie between us and the Sun, Mercury and Venus, may sometimes pass in front of the solar disk. A spectacular Venus transit occurred in 2012 (Fig. 5.8), the next passage of this kind is a Mercury passage that will happen on May 9, 2016.

How would such a phenomenon be perceived from a long distance? How much light would, for example, the largest planet, Jupiter, obstruct from the Sun? The diameter of Jupiter is about ten times smaller than the Sun's, so its surface seen as a silhouette against the Sun's disk would be hundred times smaller. Thus it would obstruct 1/100 of the solar surface, diminishing its light with 1 % as long as Jupiter is in front of the disk. In turn, Earth (or Venus) is about ten times smaller than Jupiter so the corresponding weakening due to an Earth passage would be 0.01 %. Are these measurable effects? Again,

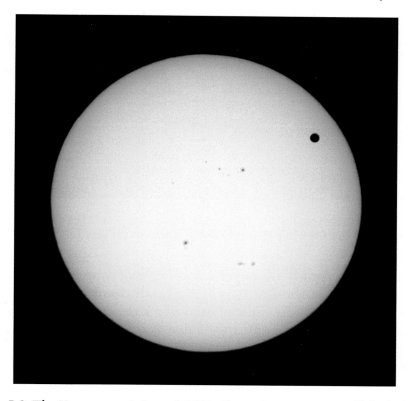

Fig. 5.8 The Venus transit June 6, 2012. Venus is seen as a *small dark disk.* The other spots are sunspots (Bengt Rosengren)

the answer is yes, in both cases. It has been possible to measure a 1 % reduction of a star's light for decades, since photoelectric photometry was introduced in the 1950s. The fact is that larger exoplanets could have been discovered already then. However, at that time there were no CCD cameras, so the stars had to be measured one by one with another kind of light detector, known as photomultiplier. But the chance to detect a Jupiter-sized planet with this technology is quite small. Discoveries using the transit method did not become realistic until the breakthrough of the panoramic and digital CCD detector, combined with smart computer algorithms. Even so, while Jupiters could be easily detected, finding Earth-sized planets remained a difficult challenge. But recently, space-based observatories have brought their detection within reach.

The principle for transit-based discoveries is founded on high precision photometry and is illustrated in Fig. 5.9. The light of the

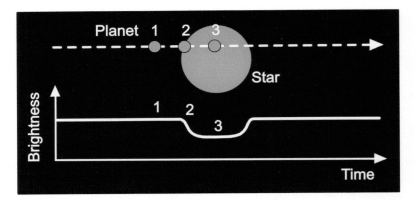

Fig. 5.9 The principle for a transit observation. An exoplanet obscures part of the stellar light when passing in front of the star (Peter Linde)

star is measured during a considerable length of time, and if a planet passes in front of the star, you see a small decrease of the light. This process is usually documented as a light curve. It should be noted that in order to confirm the discovery, the light dip should be observed several times, corresponding to the orbital period of an exoplanet. From the light curve a wealth of more detailed information can also be extracted; we will return to this in a while.

So, what is the probability that a planet has a suitable orbit, making it pass in front of its host star as seen from the Earth? It can be estimated that for a planet located as far from its star as the Earth is from the Sun, the probability is about half a percent. If the planet lies closer, the probability increases; if it is further away it decreases. But there are vast numbers of stars to examine.

Exoplanets Bend Space

A third method to discover exoplanets, successful but not so easily understood, is called gravitational microlensing. This technique takes advantage of the fact that mass affects its surrounding space. Einstein's theory of general relativity, presented in 1915, predicted this effect. The theory concerns the warping of space-time in the presence of a large mass. Just like a two-dimensional paper can be curved into a third dimension, for example a cylinder, Einstein's theory tells us that our normal three-dimensional space can also bend

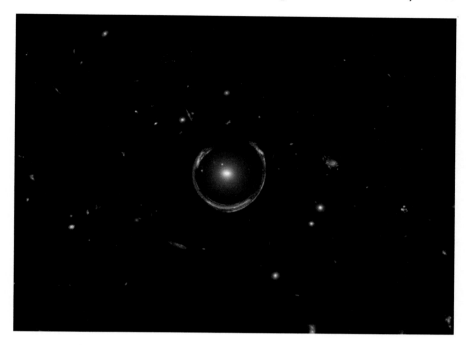

FɪG. 5.10 The cosmic horseshoe. In this remarkable image it is clearly seen
how the gravitation of the foreground galaxy curves space strongly enough
to act as a lens for the light coming from the distant background galaxy
(ESA/Hubble/NASA)

into a higher dimension. This is considerably more difficult to intui-
tively accept, but has nevertheless been proven to be true. The first
proof was delivered at the total eclipse of the Sun in 1919. During a
solar eclipse the sky darkens, and it becomes possible to see stars in
the neighbourhood of the Sun, something which is otherwise impossible from the ground. Through accurate positional measurements
of such stars, compared to corresponding measurements when the
Sun was located elsewhere on the sky, astrophysicists discovered a
discrepancy, fully corroborating Einstein's prediction. The light from
the stars had passed through a section of space curved by the gravity
of the Sun. The phenomenon was real.

The effect of the Sun on space is quite small. Across the
universe, considerably more drastic examples can be found. In
Fig. 5.10 we see the effect around a whole galaxy, about ten times
heavier than the Milky Way. It curves space so much that it works
as a de facto lens. The bright ring around the galaxy is, in fact, the
distorted image of another distant galaxy, lying very far behind! In

this case the distant galaxy lies in a straight line right behind the foreground galaxy. When its light was transmitted, the foreground galaxy was at a distance of 4.6 billion light years while the distance to the background galaxy was 11 billion light years. By amplifying the light in this way, nature gives us a unique possibility to study distant galaxies that would otherwise never have been detected.

Remarkably enough, this warping effect can also be used to discover objects as small as exoplanets. We are now back to the space curving effect from ordinary stars, like our own Sun. Our target stars in question belong to our own galaxy but may be quite distant. If two stars are exactly aligned with each other the curved space around the foreground star will act as an amplifier for the light of the background star. The most intriguing part is that the existence of an exoplanet belonging to the foreground star will also affect the space curvature enough to further amplify the effect. This subtle amplification offers another sophisticated technique for finding exoplanets. The natural movement of the two participating stars means that they will move with respect to each other, also along our line of sight to them. From a distance the phenomenon is simply observed as a short-lived increase in the brightness of a star. If you look in detail at how the brightening takes place, it is possible to reveal the existence of an exoplanet. Figure 5.11 illustrates the principle. The lower part of the figure shows schematically how the "lensing" star and its planet affects space so that the light of the distant star varies during the passage. The upper part shows the light curve for the process. The two extra peaks are due to the exoplanet.

Unlike transit observations, we are not dealing with subtle measurements of minute variations. Powerful lensing events may cause amplification between five and fifty times the original brightness of the background star, which make these events rather easy to observe when they happen. The entire light amplification process normally lasts between ten and a hundred days, but if an exoplanet is present, the change in brightness can happen on a time scale of just a few hours.

The actual implementation of this type of observation is very difficult. Stars seen from vast distances appear extremely small, and it is necessary that they lie in an exact line with respect to Earth in order to allow anything to be observed. An approximate

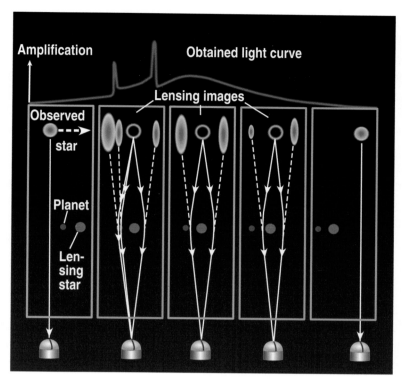

Fig. 5.11 An example of gravitational micro-lensing. The gravitation from the nearest star and its planet curves space so that the light from the background star varies (NASA/Peter Linde)

comparison would be to place a table-tennis ball at the distance of Mars and exactly behind it a tennis ball at the distance of Jupiter...

In order to have a reasonable chance to observe the phenomenon, the searches have to be made in very dense stellar fields, preferably in the direction of the centre of the Milky Way, where there are millions of stars. Another, even more extreme possibility is to use nearby galaxies like the Magellanic Clouds or the Andromeda galaxy as background. It is estimated that even where the stellar density is very high, the probability for a lensing event is less than one in one hundred million. To compensate for this it is necessary to continuously and simultaneously (and automatically!) measure the brightness of huge numbers of stars in order to catch an ongoing lensing event. In spite of these difficulties there are a number of cases where astronomers have been successful. Discovering exoplanets by the use of gravitational microlensing involves some unique properties compared to the other methods. A disadvantage is that every microlensing is a one-time event; after it has happened it never repeats. Additionally, discoveries made using this technique usually cannot be followed up by other types of supporting observations, since the distance to these stars (and their planets) is often quite large, perhaps more than 10,000 light years. But there are advantages as well. The measurements are not dependent on the brightness of the target star but on that of the background star. The target does not even have to be visible, which means that you can study planets around any type of star, independent of its intrinsic properties. For example, it is quite possible to find planets around small and faint stars. Contrary to the radial velocity measurement technique, you do not need to wait for an entire orbit before you can determine the existence of a planet. Yet another bonus is that the sensitivity of the method favours exoplanets whose distance from their star is between 0.6 and 1.6 AU. This is quite convenient since stars similar to the Sun or slightly smaller have their habitable zone approximately at this distance. The method sometimes also allows for the detection of more than one planet in the same microlensing effect.

Timing Pulsars: An Indirect Doppler Technique

Actually, a couple of observations of suspected exoplanets were made already before Michel Mayor and Didier Queloz published their discovery in 1995. Regular variations had been found in the radio pulses received from certain pulsars. A pulsar emits a distinct radio pulse once per rotation, much like a lighthouse emits light on Earth. The rotation periods of most pulsars are in the order of 0.1–3 s, and the interval between two radio pulses can be measured with extremely high precision. As in the case of normal stars, the position of a pulsar is influenced by the existence of a planet; it waggles a little. And as is the case with Doppler effects, the movement can be detected, although in this instance by means of studying the regularity of the radio pulses. If the pulsar is moving away, the pulse are delayed; if it is moving towards us they come a little bit too early. In a sense, this is also a kind of Doppler effect, measured in time instead of by the wavelength of light. The method is very sensitive.

Exoplanets in orbits around pulsars are, however, less interesting as potential bearers of life. A pulsar is really a neutron star and has formed as an extremely dense remnant after the original star annihilated itself in a gigantic supernova explosion. Thus life on any planet belonging to such stars would very probably have been erased.

Direct Observations Are Difficult

One would imagine that direct observation of an exoplanet would be the most natural and straightforward way to discover them. In fact, it is one of the most difficult observations. Since the exoplanet only reflects the light from its host star, it is much fainter and is extremely difficult to see in the glare of the star. If we once again take our Earth and Sun as examples, we find that if this system is observed from a long distance, the Earth would be more than a billion times fainter than the Sun. Additionally, Earth would appear to lie extremely close to the Sun.

To distinguish an exoplanet from its host star under such circumstances is a great challenge. However, researchers have succeeded in several cases, using very large and specially equipped telescopes. Those cases are restricted to very large planets, usually larger than Jupiter, located at large distances from their host stars. However, an arsenal of high-tech tricks can be used to facilitate the observations. These include high image resolution using adaptive optics, dampening the light of the host star by coronographic techniques, and resorting to observations further into infrared wavelengths of the light. We will return to examine these technologies in a while.

Investigating Exoplanets Is Not Transparent

There is a difference between merely discovering an exoplanet and actually studying its characteristics. We not only want to expand our postage stamp collection, we also want to determine the individual properties, to characterise and catalogue them. That phase of the exploration has recently taken its first faltering steps, and here it is only the major players that can contribute. To be able to move forward in this research, several large ventures in building large telescopes are in progress. These instruments will be located both in space and on the ground.

As we have already indicated, there is a host of complicating factors to take into account and compensate for when you try to observe exoplanets. For ground-based observations the light has to pass through the atmosphere. The turbulence in various atmospheric layers impedes exact measurements. This turbulence is obvious to anyone watching the stars in the sky on a dark night. The stars twinkle, but this not an intrinsic property of the stars themselves; the twinkling depends instead on how the light from the stars is randomly disturbed during its passage through the atmosphere. The reason is that the air, which is always in motion, has small differences in temperatures. These temperature gradients in turn create small volumes of different densities, since temperature and density are interconnected in a gas. But different density means that the optical properties of the gas are changed locally. Various small volumes will have differing refraction indices. Together with the random motions, this causes the light from an originally point-like object like a star to be spread in a way that the eye perceives as a twinkling. A physicist would call this scintillation. A corresponding (albeit simplified) version of this phenomenon takes place in shallow waters along a summer beach. At the sandy bottom small water waves will focus the sunlight in random ways, and a crab studying the sky would see a similar effect for the Sun. Another manifestation of the phenomenon is seen in the air above a working bread toaster. You can see the hot, turbulent, air moving and even casting a shadow.

In the telescope, the effects of atmospheric turbulence cause blurring, which affects and limits the achievable image resolution. However, the varying nature of the turbulence also allows for short moments of clarity, something that can be utilised with modern technology. In order to maximise the moments of good seeing, astronomers prefer to build their observatories at high altitudes, usually at exclusive and isolated places at least 2000 m above sea level. Naturally, the optimal way of avoiding atmospheric problems is to place the telescope in space. However, ground-based observatories increasingly utilise new high-tech compensating equipment, called adaptive optics, allowing for diminishing or even totally cancelling out the turbulence effects (see Box 7.2).

The atmosphere also works as a filter. Figure 5.12 gives an overview of the transparency of the atmosphere. Only certain

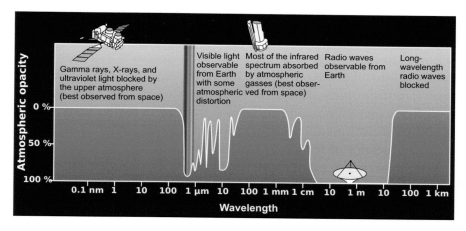

FIG. 5.12 The transparency of the atmosphere at sea levels for various wavelengths (NASA/Peter Linde)

wavelengths come through: visible light is such a case, as are radio waves. Fortunately, harmful wavelengths are blocked, such as ultraviolet and X-ray radiation. On the other hand the infrared, which is of great interest to exoplanet researchers, is also blocked. In addition to this, the laws of nature put down their own limits. It so happens that light itself has a statistical nature, which means that an exact measurement of even a constant light source is, in principle, impossible. This is due to its quantum-mechanical formation as individual photons, and is governed by the laws of atomic physics. Without going into too many details, this means that the uncertainty that is inherent in a light measurement can be estimated mathematically by taking the square root of the number of photons registered. In everyday life this is not a problem since ordinary daily light contains trillions of photons. But with modern digital cameras, the effect is actually noticeable in that dark areas of images are more noisy. The light from the stars are, however, incredibly faint. In fact, you can often count the number of single photons received from a star. A hundred photons are enough to detect a star, but in such a case the error in the brightness measurement will be at least 10 %. However, for a star with a million registered photons the error is only 0.1 %. An accurate observation of a star is thus facilitated if it is bright, if the telescope is large (capable of collecting many photons) and if it is possible to observe for a prolonged time (accumulating many photons).

Additionally, nature itself creates many additional complications. Some stars, especially the red dwarfs (also see Box 6.4), are believed to have large star spots on their surfaces, perhaps larger than the sunspots on our Sun. Of course, this affects their brightness. The light emission from the stars themselves may be variable. Stars can physically pulsate, resulting in rather large, often periodic, variations in their brightness. On the other hand, such stars are usually of lesser interest in terms of exoplanet research, at least in terms of planets suitable for life. For this we need planets located at calm and stable stars.

Double stars are common and often complicate the search for exoplanets. When seen side-on, they can pass in front of each other and obscure each other's light, forming what is known as eclipsing variables. In the hunt for small brightness variations, many more eclipsing variables than exoplanets have been found, especially when utilising the transit method.

Even if a measurement becomes accurate enough, it takes time before successive observations have been performed, verifying the existence of an exoplanet. If we once again take our own solar system as a reference, it would mean that, for an outside observer, it will take a year between each time the Earth passes in front of the Sun. For the case of Jupiter, 12 years are needed for a second observation. Similar time delays are necessary for measurements based on radial velocities.

Applying Sophisticated Techniques

While the techniques to extract knowledge from the light originating from the exoplanets and their host stars are already amazingly sophisticated, they are continuously being improved. Transit observations provide a wealth of information. The brightness reduction gives a direct measure of the size of the planet compared to the host star, and since the size of the host star is often known from other observations, this gives a measure of the planet's real size.

But if we can achieve very high accuracy in the photometric measurements, even more can be deduced. Since we are looking at the system sideways for transit observations, the planet will be illuminated by its star in such a way that it shows phases, similar

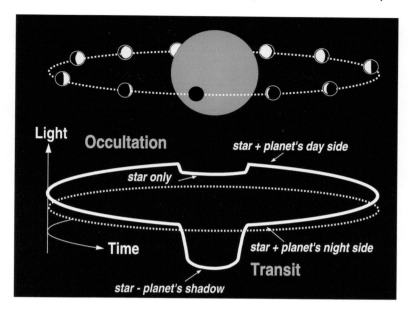

FIG. 5.13 High precision measurements allow for an even more detailed light curve than shown in Fig. 5.9. The motion of the planet affects the total brightness in various ways and this also gives measurable effects in the spectrum of the light (Peter Linde after an original by Michael Perryman)

to how the Moon appears in our sky. It has already become possible to study these variations in the combined light of the star and its planet, at least for giant Jupiter-sized ones (see Fig. 5.13).

Additionally, information about a possible planetary atmosphere can be hidden in the light that reaches the Earth during the eclipse. Let us assume that the exoplanet has an atmosphere. Light from the star will pass through the atmosphere of the planet on its way towards Earth and thus be modified a little by the elements in planetary atmosphere (see also Fig. 6.26). This is a spectroscopically observable effect, although quite small. As mentioned above, the sum of the light from the exoplanet and its host star can also be studied. The composition of the atmosphere affects the reflected light from the exoplanet, and even if its light nearly drowns in the light of the host star, this effect can be studied, supplying further information about the atmosphere of the planet.

The ingenuity of the scientists just continues. As another example, it is now even possible to tell whether an exoplanet is moving around its host star in the same direction as the host star rotates or if it moves in the counter direction.

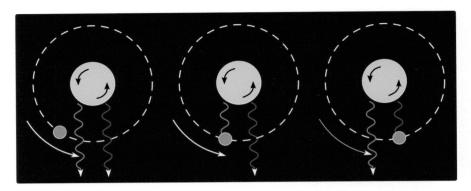

FIG. **5.14** An exoplanet passing in front of its host star will first block some light from one side of the star, then from the other side. The phenomenon reveals if the orbital direction of the planet is the same as the rotational direction of the star (Nicholas Shanks)

In Fig. 5.14 we see how scientists resolve this movement. Yet again, they utilise our favourite tool, the Doppler effect. Let us assume that we are seeing the star sideways, i.e. perpendicular to the rotation axis of the star. In the figure, this direction is downwards. Keep in mind that the only thing we can actually observe is the light from point-like star that enters a telescope and is analysed in an advanced spectrograph. Due to the star's rotation, a fraction of its light comes from that half that is moving towards us, and this light becomes slightly blue-shifted. The other half is receding from us, and that light correspondingly becomes red-shifted. Again, this is a very small effect, but for some stars it is nevertheless measurable. Now imagine that an exoplanet in transit passes through the line of sight. It will cover a part of the stellar light, and if it is moving in the same direction as the rotation of the star it will first obscure some of the blue-shifted, then subsequently some of red-shifted light. These measurements are so sensitive that even such minute effects on the light of the stars is occasionally observable. The phenomenon is known as the Rossiter-McLaughlin effect.

At the same time it must also be remembered that there are many sources of errors. We have already discussed the problem with aberrations of the incoming light. But there are other problems of a more technical nature. The instruments are usually pressed to their limits, sometimes even beyond the original specifications. A very meticulous and systematic checking of mechanics and electronics of the instruments must be performed. Their properties must be examined in detail and a thorough calibration is of utmost importance in order to minimise various kinds of error sources.

Who Studies the Exoplanets?

Since 1995 a large number of exoplanets have been discovered, and organisations now direct substantial resources toward this kind of research. Move than a hundred projects all over the world are operating or in the planning stage. In Table 5.1 the most important projects with a record of successful discoveries are listed, sorted by the number of discoveries. The numbers are approximate but give a good overview of which projects and instruments that so far have been successful in exoplanet research. Until 2010, most of the discoveries were made from the ground, but since then a couple of specialised satellites have made very significant contributions. The first discoveries were made using the radial velocity method, which is still very important. However, discoveries using the transit method are now dominating, and this trend will continue. At present, (2015) about 60 % of the discoveries have been made with this method, with about 30 % using the radial velocity technique. The rest have been discovered with the other techniques mentioned earlier. The different techniques are usually complimentary, resulting in more information, if they can be combined for the same exoplanet.

So, using modern technology, it is rather easy to discover large exoplanets. Even registering a brightness decrease of about 1 % using the transit method can be accomplished with rather modest equipment. For this reason, a number of ongoing low-cost projects have been successful in competing with the big elephants. Additionally, such projects are usually tailor-made for exoplanet studies, carrying out very long sequences of observations. This is invaluable for this type of research. In contrast, astronomers use the telescopes at the largest observatories for many types of astronomical research, all competing for observing time. Small exoplanets are much more difficult to discover and require quite

TABLE 5.1 Some of the most important observing projects which have resulted in discoveries of exoplanets. Size of telescope is given as well as designation of spectrograph used

Project	Method	Discoveries	Place	Comments
Kepler 0.95 m	Transit	1000+	NASA, Space-based	Satellite in orbit around the Sun
Keck 10 m/HIRES	Radial velocity	154	Hawaii, USA	Spectroscopy based
ESO 3.6 m/HARPS	Radial velocity	130	ESO, La Silla, Chile	Spectroscopy based
SuperWASP, 16×0.2 m	Transit	92	South Africa, Canary Islands	Survey of bright stars
HATnet 3×0.11 m+24×0.18 m	Transit	58	Several places in the world	Survey of bright stars
Euler 1.2 m/Coralie	Radial velocity	39?	ESO, La Silla, Chile	Spectroscopy based
AAT 3.9 m/UCLES	Radial velocity	28	Australia	Spectroscopy based
CoRoT 0.27 m	Transit	28	ESA, space-based	Satellit in orbit around Earth
OGLE 1.3 m	Microlensing, transit	26	Las Campanas, Chile	Also dark matter studies
HET 9.2 m/HRS	Radial velocity	25	Texas, USA	Spectroscopy based
OHP 1.93 m/Elodie	Radial velocity	18	France	Spectroscopy based
MOA 1.8 m	Microlensing	14	New Zeeland, Japan	
Okayama/HIDES	Radial velocity	12?	Japan	Spectroscopy based
Magellan 6.5 m/MIKE	Radial velocity	10	Las Campanas, Chile	Spectroscopy based
Tautenburg	Radial velocity	7	Germany	Spectroscopy based
XO	Transit	7	Hawaii, North America, Europe	Survey of bright stars, amateur collaboration
Lick/Hamilton	Radial velocity	5?	California, USA	Spectroscopy based
OHP 1.93 m/Sophie	Radial velocity	5	France	Spectroscopy based
TrES 3×0.1 m	Transit	5	California, USA, Canary Islands	Survey of bright stars

different resources. In conclusion, many types of telescopes are useful in exoplanet studies, both small and large, ground-based or space-based. Even amateur astronomers and the general public can help with the research. In fact, all sorts of observations are needed and support each other.

Hence, smaller telescopes are primarily capable of detecting large Jupiter-sized planets using the transit technique and massive studies of in large stellar fields with millions of stars. This is nevertheless quite important for focusing attention to these systems, since it is assumed that systems with giant planets should contain also smaller planets. Larger telescopes can subsequently study them in more detail. As more and more Earth-like planets successively are found, they will of course get the full attention of the very largest ground-based and the space-based telescopes. Now let us have a closer look at some of the players in this race.

Reconnaissance From The Ground: Small Is Beautiful

By using small and ground-based (as well as cheaper) telescopes, it is possible to contribute to the exoplanet research. Such facilities normally use photoelectric techniques with panoramic detectors to scan through the sky, looking for tiny brightness variations. A few are geared for microlensing events, but most rely on the transit method. This type of work requires the capacity to accurately and automatically measure the light of large numbers of stars. And a lot of patience. Only a small fraction of the presumptive exoplanet systems will have their orbital plane pointing towards the Earth, and a huge amount of measurements need to be done, both of many stars and at short time intervals. A typical planet passage lasts between 2 and 8 h. The goal is to get as many observations as possible during this time. But between events, observers may have to wait from days up to several years before the next passage occurs. Automatic systems with fast and accurate measurements are therefore necessary. Fortunately, this technology is now readily available. Modern CCD-detectors are extremely light-sensitive and accurate (see Box 5.2). In addition, they are relatively cheap, and the digital handling means that the necessary calibra-

Fig. 5.15 The XO-instrument contains two 20 cm telephoto lenses, each with its CCD camera (NASA/ESA /J. Stys/STSci)

tion can be made with high precision. This is a prerequisite for the subsequent measuring phase where advanced programs search the images and measure the brightnesses of the stars again and again, looking for slight changes. Several ongoing projects with smaller telescopes successfully utilise these methods.

A project called XO uses an instrument that resembles a large binocular (Fig. 5.15). It is located at an altitude of 3000 m at the peak of the Haleakala volcano on Hawaii. Data from thousands of stars is analysed for traces of light variations. The most interesting candidates are forwarded to a network of amateur astronomers in North America and Europe for detailed studies. When the existence of an exoplanet has been confirmed, a larger observatory supplements the identification with radial velocity observations. So far, XO has been able to find five Jupiter type exoplanets.

A similar project is HATnet (Hungarian Automated Telescope Network) which also consists of a set of smaller instruments. The prototype was developed in Hungary and was originally nothing more than a modest telephoto lens with a 65 mm aperture, connected to a small CCD camera. The system has progressed, with several 11 cm telescopes now located at smaller observatories, spread out over the Northern hemisphere. The advantage of such diverse locations is that it affords good opportunities to cover the whole course of a transit. When the object in question sets below the horizon at one observational site, another one, located more to the East, can take over. However, from locations at the Northern hemisphere the Southern part can never be observed. Consequently, a HAT-South was established in 2009, with instruments in Namibia, Chile and Australia. So far, HATnet has detected about 60 exoplanets in total.

The Polish project OGLE (Optical Gravitational Lensing Experiment) started in 1992 by looking for microlensing effects, and since 2001 it has also looked for transits. As we know, both methods rely on accurate light measurements and can be combined for better data. OGLE uses a medium-sized 1.3 m telescope at the Las Campanas observatory in Chile for the observations. So far, 27 exoplanets are on their record.

Perhaps the most ambitious project for ground-based surveys is WASP (Wide Angle Search for Planets), recently upgraded to SuperWASP. The project is a collaboration between British universities and uses two instruments, one at the Canary Islands and one in South Africa. Each instrument contains eight powerful telephoto lenses with CCD cameras connected to them. The combined field of view is very large, monitoring hundreds of thousands of stars simultaneously. The project has already led to the discovery of about 100 Jupiter-sized planets.

There is also room for even smaller but still high-tech ventures in the realm of exoplanet research. Quatar has a telescope located in New Mexico, USA, which has made two discoveries. Probably the smallest of all instruments used is in the KELT-project. A 4 cm (!) telescope with a very large field (essentially a telephoto lens with 80 mm focal length) has been specialised for studies of stars that are actually too bright for observing with larger telescopes. At least one confirmed exoplanet detection has been reported.

A number of projects are also active on the microlensing front. The OGLE-project regularly monitors about 200 million stars. A similar Japanese-New Zealand collaboration with the acronym MOA (Microlensing Observations in Astrophysics) is carrying out observations from New Zealand. Both these groups perform comprehensive surveys in the hunt for microlensing events. When a suspected event is registered, an alert is sent to a number collaborating groups worldwide. A microlensing event normally occurs over the course of a few weeks, leaving time to start observational campaigns in order to closely examine the event. In this way the accuracy in the measurements increases as does the possibility of covering the entire time process, independent of location on Earth. There are a number of cooperating smaller observatories, including advanced amateur astronomers, that lend a hand in this work. Some observatories are completely automatic and robotically controlled. Following an alert, these can very quickly make supportive observations.

The Big Elephants

Naturally, many participate in the hunt for exoplanets. The bigger observatories in the world made the first discoveries and their efforts still usually are in the front line. But larger observatories are needed to make more demanding discoveries and for characterisation of exoplanets. In addition to large telescopes, the search requires specialised instruments. As we pointed out in the beginning of this chapter, to successfully utilise the radial velocity method, very advanced spectrographs are needed in order to detect the exoplanets as well as determine their orbits. Instruments of this calibre are quite rare, although several new ones are under development.

As noted previously, the breakthrough in the hunt for exoplanets came with Michel Mayor's and Didier Queloz's discovery of the planet around 51 Pegasi. Mayor and his Swiss-French group used a 2-m telescope in the French Pyrenees for their observations. Almost simultaneously, Geoff Marcy and his group at the Lick Observatory in California reported the first American exoplanet discoveries. After this, research accelerated rapidly with further improvements made on spectrography technology. Since then, many European discoveries have been made by the ESO observatory at La Silla in Chile. The big 3.6 m telescope, the first large European telescope in the Southern hemisphere, has, to a large extent, been dedicated to exoplanet studies. Here, ESO has installed the world leading HARPS spectrograph. In addition, the smaller 1.2 m Swiss Euler telescope, also at La Silla, has been equipped with an excellent spectrograph. Together, these two telescopes have generated about 170 discoveries.

In the USA, most of the radial velocity-based discoveries have been made with the spectrograph HIRES attached to the Keck I

Box 5.2: How Does a CCD-Camera Work?

CCD (Charged-Coupled Device) cameras have completely dominated astronomy since the 1980s as panoramic detectors. Their light sensitivity and their capability of digital handling have led to a new epoch of astronomical discoveries. Today a variant of them is also used in ordinary consumer cameras.

The principle for a CCD is illustrated in Fig. 5.16. A number of pixels arranged in a rectangle are used as collecting elements for the incoming light. The pixels are very light-sensitive semiconductors, made from silicon, which can register up to about 90 % of the infalling light. The CCD collects light during a fixed time (exposure time) and the light (photons) creates an electrical charge in the form of electrons that successively are stored and trapped in each illuminated pixel. After a finished exposure, the charge in the pixels is shifted out row by row. Each pixel charge is transferred systematically into a special column readout register. This, in turn, shifts each charge out of the CCD. In the process, the weak electrical charge is amplified, measured and digitised. For scientific purposes, the CCD also needs to be cooled in order to avoid the spontaneous creation of electrons via thermal effects (so called dark current).

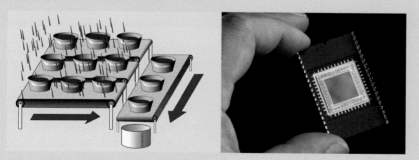

FIG. **5.16** *Left*: The basic principle for a CCD camera (simplified). *Right*: A CCD chip for astronomical applications (Peter Linde, NASA)

Improved, larger CCD chips have now been developed to a size of at least 4096×4096 pixels, i.e. 16 Mpixels. By putting several together in a mosaic, the receiving surface can be made even larger. The largest camera in astronomy, now under development, will have 3.2 Gpixels, i.e. more than three billion pixels.

Fig. 5.17 The American huge telescopes Keck I and Keck II are located at an altitude of more than 4000 m at the summit of Mauna Kea on Hawaii (Science Photo Library/IBL)

telescope at Hawaii (Fig. 5.17). The Keck I stands as one of the world's largest telescopes, with a segmented mirror with a total diameter of 10 m. More than 150 discoveries can be ascribed to the group associated with the Keck telescope. Table 5.1 displays some additional large observatories contributing to exoplanets studies, including the Hobby-Eberly telescope in Texas and the Anglo-Australian Telescope at Siding Springs in Australia.

A large number of medium-sized telescopes around the world—some threatened to be shutdown for economical reasons—have experienced a second life in the hunt for exoplanets. Among these, several networks have been established to make the search more effective.

Space Telescopes on the Hunt

Space-based observations have brought astronomy forward in great leaps. New windows to the universe have opened up by their orbital observations in wavelengths which cannot penetrate to the ground. Above all, progress have been made in observing shorter wavelength radiation like ultraviolet and X-ray, but also longer wavelengths such as infrared. Figure 5.12 clarifies this. Observations at these wavelengths have resulted in many discoveries that have deepened our knowledge about the universe. Many space telescopes are highly specialised, but not all. Most famous of the non-specialised space telescopes is the Hubble Space Telescope, primarily intended for observations in visible light.

For exoplanet studies, the foremost advantage with space-based observations is the absence of an interfering atmosphere. Most types of measurements are based on some sort of advanced optics, but the air and its turbulence has a negative impact. Conditions in space open up the potential of extremely accurate measurements of the brightness and positions of the stars. These are critically important aspects of exoplanet research. Space-based observatories enjoy another great advantage: undisturbed observations in the infrared. This suits exoplanet observations perfectly, since the contrast between host star and exoplanet is considerably lower at these infrared wavelengths (see Fig. 6.3). The large Spitzer telescope, launched in 2003, is optimised for observations in the infrared. The telescope itself has not discovered any exoplanets but has played an important (and unexpected!) role in finding clues to the composition of the atmospheres of exoplanets.

Engineers planned the Hubble telescope in the 1970s, but delays—in particular due to the catastrophe of the space shuttle Challenger, which exploded shortly after launch in 1986—meant that it did not enter space until 1990. The first images obtained from space were a great disappointment. As mentioned, the telescope mirror had been polished to the wrong shape. Thus Hubble became fully operative about ten years later than planned. The built-in serviceability, however, immediately proved very useful. Exoplanets were hardly contemplated when Hubble became operative, but the telescope has since played a role in this type of research. However, like the major ground-based observatories, Hubble observation time cannot exclusively be used to look only for exoplanets. But some important studies have been made. For a few days in 2000, controllers pointed the Hubble telescope toward a very dense star cluster. Expectations were high to find exoplanets. The result was a disappointment: not a single one could be verified. This indicates that the conditions inside such a concentrated cluster may not be suitable for stable planetary systems to form. A few years later, a large project with the name SWEEPS (Sagittarius Window Eclipsing Extrasolar Planet Search) was conducted. Astronomers selected a very crowded stellar field in the direction of the Milky Way centre, about 27,000 light years away. Observations were performed over the course of a week, and accurate measurements were made for 180,000 stars. Only 16 candidates were noted, of which two have been verified as exoplanets so far. However, the Hubble telescope has been of assistance doing detailed studies of exoplanets already detected in other, ground-based, projects.

The first satellite launched in order to specially search for exoplanets was the European COROT satellite. During 2006–2014 this French built satellite was operational. A small 27 cm telescope collected the light. Researchers searched for both exoplanets and for seismic phenomena in stars. This successful satellite discovered 28 exoplanets, along with another 600 suspected candidates. However, in 2009 COROT became overshadowed by an even more advanced satellite.

The Kepler Project

How common are Earths in our galaxy? How many Earth-like planets orbit inside the habitable zones of solar-type stars? These ambitious but fundamental questions were the starting points when designers planned the Kepler telescope. Like most satellite

Fɪɢ. **5.18** The Kepler telescope (NASA/Ames/JPL-Caltech)

projects, Kepler has a long history. In 1992 the first plans were presented, but NASA dismissed them as unrealistic. It was not until
the fourth attempt, in December 2001, that NASA finally approved
the project. During the intervening nine years the technological
prerequisites had improved enormously. Now CCD cameras with
the necessary specifications were available, as was the computing power necessary for the demanding and complicated analysis
of the observations. Meanwhile, the first ground-based discoveries of exoplanets had been made, something which substantially
increased the status of the project. After a variety of further delays,
among them economics, the satellite finally launched in March
2009 (see Fig. 5.18).

The Kepler satellite soon became the dominating player in
the exoplanet race. At regular intervals, the Kepler team continues
to announce new discoveries from this unique observatory, which
is specially designed to use the transit method to find exoplanets.
Let us have a closer look at Kepler and its mode of operation.

The satellite itself is rather simple. It contains what is known as
a Schmidt telescope with an aperture of 95 cm. Similar telescopes
have been used for more than fifty years in ground-based observa-

tions, and their principal property is to give a large field of view combined with high imaging quality. In the case of the Kepler satellite the field of view is 115 square degrees. This approximately corresponds to the field covered by a fist, held at arm's length. One of the technological challenges was to fill this large field with a sensitive detector. Engineers found the solution in a composite camera consisting of 42 CCDs, each with 2220×1024 pixels, altogether close to 100 million pixels!

The satellite was positioned in a very special orbit. It is heliocentric; the satellite orbits the Sun just like Earth does. The orbit is very similar to Earth's but its period around the Sun is 372.5 days, i.e. about seven days longer than Earth's. This means that it successively falls behind, each month receding from Earth more than a million kilometres. This orbit was deemed the most suitable for several reasons, among them to save on manoeuvering needs. Also, since it is far away from the Earth, the telescope does not suffer any disturbances from here—neither Earth light, gravitation, magnetic fields nor atmosphere can cause any problems. On the other hand, it still remains sensitive to radiation outbursts from the Sun.

Is the satellite really capable of finding exoplanets the size of Earth? As we found in the beginning of this chapter, the brightness decrease we need to detect is very small indeed, about 0.01 % of the light of the host star.. Another way to express this is in ppm (parts per million). 0.01 % is equivalent to 100 ppm. The Kepler telescope is designed to detect differences as small as 20 ppm for the stars of interest, and thus should be capable of finding planets the size of Earth. To get a feeling for this kind of accuracy, one can imagine two exactly identical lampposts seen at a distance of 1 km. 20 ppm corresponds to the difference in the received light arising from moving one of the lampposts 1 cm further away!

The working principle of the satellite is quite simple, at least in theory. It continuously observes the same area of the sky, a field that lies high in the Northern sky, between the constellations Cygnus and Lyra (Fig. 5.19). The field is selected to give the best possible results. The direction is such that the satellite observes along the line which the solar system itself is moving in the Milky Way. It observes stars that lie approximately at the same distance from

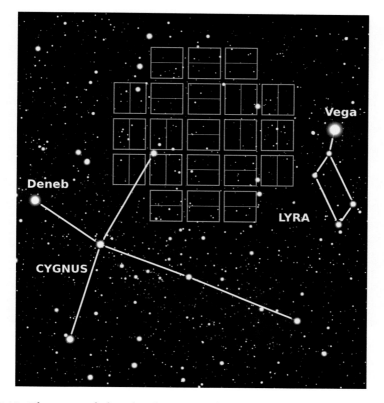

FIG. 5.19 The part of the sky, between the constellations of Cygnus and Lyra, surveyed by the Kepler satellite (Peter Linde)

the galaxy centre as the Sun does, and like the Sun, they are located close to the symmetry plane of the galaxy. Both these factors mean that Kepler observes stars located in what may be considered the habitable zone of the galaxy (see Chap. 10). Figure 5.20 shows data from the upper part of the Kepler field. Hundreds of thousands of stars are being monitored simultaneously.

The observations follow a simple routine but the data analysis is very advanced. Every sixth second the 100 million pixels are read out and the information is addititively combined to a total exposure time of 30 min. A considerable amount of triage analysis is carried out, removing very faints objects, etc. About once a month data are then transmitted to Earth for further analysis.

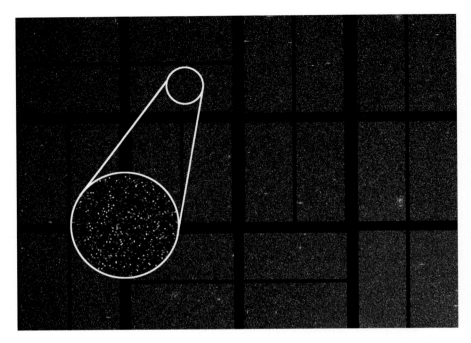

Fig. 5.20 Part of the Kepler view of field. Hundreds of thousands of stars are monitored at the same time (NASA/Ames/JPL-Caltech)

The observations made by the Kepler satellite have been very successful. As of November 2015, it has discovered 1030 new exoplanets, in addition to which there are a further 4694 candidates. From these it is estimated that at least 250 may be the size of Earth. Many of these will surely be confirmed in the near future. Candidates are normally confirmed by special studies using the radial velocity method from some ground-based observatory. It should be mentioned that in 2013, Kepler became handicapped due to problems with the steering control, but since May 2014 it has continued with astronomical observations of somewhat lower quality, but it will still be able to detect short-period Neptunes and super Earths. The results from Kepler and all the other discoveries will be studied in some detail in the next chapter.

6. A Menagerie of Planets

Now, let us take a look at the new results for the planets in other planetary systems. What have we learned? Do we know anything new that we did not already know from our own solar system? The answer is a big YES.

Until 1995 the recipe for a well ordered planetary system was reasonably well established. It was believed that we held a good understanding of how a planetary system develops, although many of the details were still missing. The classical picture contained a dominating star at the centre of the planetary system. Circling the star were a few smaller rocky planets, with some large gas giants further out, and further still some smaller gas ice giants. All orbited the host star in the same plane, with stable, almost circular orbit shapes. Some remaining material, in the form of asteroids and a more distant surrounding of comet embryos and comets, completed the picture. The theories behind such a scenario were also well established; we have previously presented the fundamental arguments for these. Unfortunately (or perhaps fortunately) much of this now needs to be revised, complimented and extended. To base theories on a single observable example, our own solar system, is of course quite shaky, a fact which astronomers themselves are the first to admit.

When new discoveries of exoplanets began to pour in, everything immediately became much more complicated. The old picture turned out to be rather simplified. Of course the simplest discoveries—large and heavy planets—came first. Those who moved the fastest around their star were discovered first, since only a short time was needed to verify them. Nevertheless, it came as a great surprise that the first discoveries revealed completely different types of planetary systems. Apparently there were a lot of giant planets that had orbits very close to their star. Discoveries until 2000 were dominated by this new type of planets. They were quickly dubbed "Hot Jupiters". Michel Mayor's and Didier Queloz's first historic announcement concerned such a case.

© Springer International Publishing Switzerland 2016
P. Linde, *The Hunt for Alien Life*, Astronomers' Universe,
DOI 10.1007/978-3-319-24118-0_6

The star was 51 Pegasi in the constellation of Pegasus, a quite ordinary star that resembles our own Sun. But the planet in question was truly remarkable. It's remarkable nature did not have to do with its mass—it was estimated at half a Jupiter mass—but the fact that the distance to the host star was only 8 million kilometres. Compare this to the distance of the Earth to the Sun, 150 million kilometres, or to our innermost planet, little Mercury, which orbits 58 million kilometres from the Sun. It takes Mercury 88 days to complete an orbit around the Sun. In the case of 51 Peg, the planet completed an orbit in four days. That meant that the radiation influx on the planet was very high, perhaps 400 times higher than on Mercury, which in turn pointed to very high temperatures on the planet, perhaps up to 2000 °C. Mayor had obviously discovered a new type of planet—albeit not anything like Earth. An artist's conception of how this scenario may look like is seen in Fig. 6.1.

More discoveries of Hot Jupiter-type planets came thick and fast, but in the first wave of discoveries there was already news. The orbits of several of the exoplanets did not comply with the

FIG. 6.1 View from a planet orbiting very close to its star. The apparent size of the star is 20 times larger than the Sun on our sky and its brightness is hundreds of times more intensive. The temperature on such a planet can reach 2000 °C (ESO/L Calcada)

accepted view. They often were not the same as in our solar system. On the contrary, it was demonstrated that many had strongly elliptical orbits whose shortest and longest distance to the star could be very different. So, at an early stage, it became clear that planetary systems around other stars could differ very much from our own. It was high time for the theoreticians to modify and develop their understanding for how a planetary system forms. Before we delve into the exciting new data, we need some background in nomenclature and basic stellar physics, as well as a summary of important developments in exoplanet research (Boxes 6.1–6.4).

The Nomenclature for Exoplanets

After the avalanche of exoplanet discoveries, nearly 2000 at the beginning of 2015, there is quite a wealth of information to digest. As we have already hinted, the observations comprise an abundance of surprises. But let us start by discussing how the exoplanets get their sometimes complicated and prosaic designations. Some examples are 55 Cnc b, 7 CMA b and HD 60532 c. In astronomy the International Astronomical Union (IAU) sets the standard for how the naming of various celestial objects should be done. They are constantly haunted by self-appointed name givers, often in the form of commercial enterprises who sell "exclusive" rights to everything from lunar properties to stars and exoplanets. It is well worth quoting from the web page (www.iau.org/public) of the normally formal and strict IAU, where they, somewhat poetically, state the rules:

"Thus, like true love and many other of the best things in human life, the beauty of the night sky is not for sale, but is free for all to enjoy. True, the 'gift' of a star may open someone's eyes to the beauty of the night sky. This is indeed a worthy goal, but it does not justify deceiving people into believing that real star names can be bought like any other commodity. Despite some misleading hype several companies compete in this business, both nationally and internationally. And already in our own Milky Way there may be millions of stars with planets whose inhabitants have equal or better rights than we to name 'their' star, just as humans have done with the Sun (which of course itself has different names in different languages)."

Unofficial names of exoplanets sometimes circulate in the media (e.g. Osiris, Bellerophon, Zarmina and Methusaleh) but they are not recognised by the IAU. However, recently plans have been announced that the general public via organisations will have the possibility to give name propositions that after consideration will be made official.

The current naming scheme begins with the designation of the host star, as given in some relevant and recognised catalogue. Examples are those of Flamsteed's or Gliese's (see Box 6.1). The first planet discovered at the star gets the suffix b, the next c, and so on. The order of distance to the host star only plays a role if several planets for the same star are made official at the same time, and in that case they are numbered from within outwards.

As we already noted, most of the exoplanets discovered so far are found around stars relatively close to us. In fact, about 75 of these stars are visible to the naked eye in a dark sky. In Fig. 6.2 six

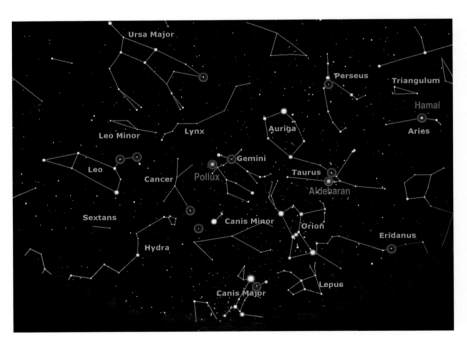

Fig. 6.2 Some of the stars visible by naked eye have exoplanets. In this map, showing a winter sky (Northern hemisphere), 13 of them are marked with *red circles* (Peter Linde/Stellarium)

of the brightest ones are marked, namely β Gem (more known as Pollux) in Gemini, α Ari (Hamal) in Aries, γ Leo A (Algieba) in Leo, ε Tau (Ain) in Taurus, ε Eri in Eridanus and 7 CMa in Ursa Major. The planets at these stars are all heavy planets of the Jupiter type, although not very close to their hosts. Smaller planets may well exist within these systems.

Box 6.1: How Do the Stars Get Their Names?

The basis for identification of objects in the sky is a coordinate system where the exact position of the object can be defined. The two celestial coordinates are called right ascension and declination, respectively, and are somewhat reminiscent of the geographical coordinates of longitude and latitude. Traditionally, only the brightest stars have proper names, often of Arabic or Latin origin. Others have designations of various types. In 1603, the German astronomer Bayer published a catalogue which assigned Greek letter designations to the stars within their constellations, according to stellar brightness. For example, the Pole Star is designated "α Ursae Minoris" (abbreviated as "α UMi"), where Ursae Minoris is the Latin genitive for the constellation Ursa Minor. The second brightest star was given the letter β, the third γ, and so on. With the introduction of the telescope, many more stars could be seen and more designations were needed. In the eighteenth century the British astronomer Flamsteed compiled a calogue with more than 2500 stars. He followed the same principle as Bayer, but used numbers instead. "61 Cygni" (star 61 in the constellation of Cygnus) is an example, "47 UMi" another.

In modern times, many catalogues have been added. The Henry Draper catalogue (225,000 stars) has HD numbers, the Hipparcos catalogue (118,000 stars) has HIP numbers and so on. Within exoplanet research, special catalogues are often used. The catalogue for Gliese contains 3800 of the nearest stars. The satellite projects COROT and Kepler have themselves designated many of their target stars.

Box 6.2: Radiation from Celestial Bodies

Every physical body emits electromagnetic radiation. This is called black-body radiation. The same laws of physics apply to an electric cooking plate as to stars and planets. When you turn on a cooking plate it gets progressively hotter. It begins emitting infrared radiation, which we perceive as heat. With rising temperature the emitted radiation develops shorter wavelengths and becomes visible, first as deep-red, then light-red and orange. This inverse relationship between temperature and wavelength is governed by Planck's law.

The surface temperature of the Sun is about 5500 °C, and the Earth's is on average about 15 °C. In the case of the Sun, its maximum intensity light is yellow, while the Earth

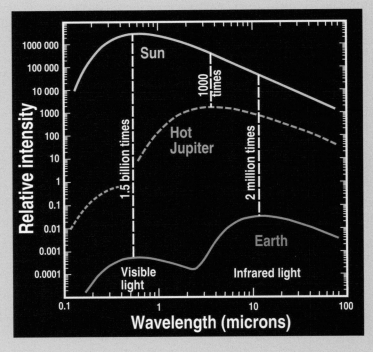

Fig. 6.3 The solar energy output, as a function of wavelength, compared to Earth's and to an exoplanet of the hot Jupiter type. The ordinate axis has logarithmic scaling, where each major tick corresponds to a change of a factor of ten. In the infrared part the Sun is less dominating (Peter Linde)

correspondingly transmits in the far infrared region, at about 10 μm, and at an immensely lower level.

Figure 6.3 shows how the radiation is distributed at different wavelengths. The intensity scale is relative and assumes that both bodies are seen from the same distance. Both axes are logarithmic, which means that the difference between two major tic marks corresponds to a factor of ten. As seen from the outside, the energy emission of Earth shows two different maxima. The first, in visible light, corresponds to reflected light from the Sun and carries the Sun's signature. The second, in infrared, corresponds the Earth's own heat emission. Also shown in the figure is also the corresponding curve for a hot Jupiter planet.

A Look at the Statistics

For those who want to follow the progress of exoplanet research in more detail there are a number of internet sites available as well as several apps for smartphones. Two excellent sites are the American www.exoplanets.org and the European www.exoplanet. eu. Both are continuously updated with the latest information. The European site is updated more frequently with respect to newly discovered exoplanets, while the American focuses more on completeness. It should, however, be stressed that conclusive data is often marginal and there might well be different opinions as to whether an exoplanet has been confirmed or not. Here we use data from www.exoplanet.eu.

To try to summarise what we know today about the exoplanets is in itself a rather risky undertaking. We are in the middle of ongoing developments, where new results are presented more or less daily. The truth of today can easily be obsolete tomorrow. On the other hand, as we have seen, speculations of yesterday may rapidly become the reality of today. Currently (Nov 2015), the score board looks like this: 2003 planets are acknowledged in 1268 different planetary systems. In 498 of these more than one planet has been discovered. 198 doubtful cases exist as well as a large group with more than 4700 candidate planets. A short history of the most important advancements in exoplanet research is given in Box 6.3.

Box 6.3: Important Milestones in Exoplanet Research

1995	First discovery of a planet orbiting a solar-type star. It is shown to be a hot Jupiter
1998	First discovery of a planetary system with several planets
1999	First observation of an exoplanet using the transit method
2001	Using the Hubble space telescope, a substance—sodium—is first identified in an exoplanet's atmosphere
2001	The first planet inside a star's habitable zone is found. However, it is a giant, almost 2000 times heavier than Earth
2002	A planet is found in orbit around a star 13 times heavier than the Sun, proving that planets can survive a star's evolution into a red giant
2002	First discovery of a Jupiter-type planet with a similar distance to its star as Jupiter is in our solar system
2005	Using the Spitzer telescope, the first direct observation of an exoplanet is made in infrared light
2007	Water vapour is detected in the atmosphere of an exoplanet
2008	First images of exoplanets seen in visual light
2009	The first exoplanet smaller than Earth is discovered by the Kepler satellite
2012	The first water world is discovered
2013	Discovery of first exoplanet with a density similar to Earth's
2014	Magnetic field is detected at an exoplanet
2014	Rotation period determined for an exoplanet
2015	Evidence for strong volcanism on a super Earth is found
2015	The Kepler satellite has discovered 1048 planets (June 2015) and almost 5000 candidates have been identified

Box 6.4: A Summary of Stellar Evolution

Two important observable properties of a star are its temperature and its luminosity (intrinsic brightness). If we plot these properties in a luminosity-temperature diagram, we will notice (see Fig. 6.4) that they are not randomly located, but instead are concentrated in certain areas. Such a diagram is called a Hertzsprung-Russell diagram, a chart that holds information about stellar evolution. As stars form, they quickly enter the main sequence. There, they remain stable for most of their life span. However, the initial mass of the star determines where along the main sequence it resides. The heaviest and hottest stars are located to the upper left and have a short life span, often less than 100 million years. They are called blue giants. The lightest and coolest stars lie

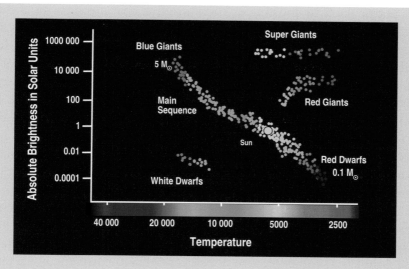

FIG. 6.4 The distribution of the stars in a simplified Herzsprung-Russel diagram. Luminosity and masses are given in solar units. The Sun's place in the diagram is marked with a circle (Peter Linde)

down to the right, having a stable life span of up to about 100 billion years. They are called red dwarfs.

When the nuclear fuel at the centres of the stars is near exhaustion, they rapidly expand into red giants and populate the upper right parts of the diagram. During this phase, they may eject considerable amounts of the stellar matter into space. Later still, the remnants collapse and, depending on remaining mass, end up either as white dwarfs, neutron stars or black holes (see also Boxes 10.1 and 11.1).

Hertzsprung-Russell diagrams are very useful tools in astronomy, especially for stellar clusters, where their interpretation is facilitated by the fact that all stars in the cluster lie approximately the same distance from Earth and have been created at approximately the same time.

Our knowledge of stars is founded on an unusually successful synthesis of observation and theory. In many ways, stars can be considered as laboratories for different kinds of physics, especially nuclear and atomic physics. As an example, today the details of how energy is created inside the stars is considered as well understood.

Let us start by looking at the statistics that are now available. The rapid pace of discoveries and the fast technological advances are illustrated in Figs. 6.5 and 6.6. Figure 6.5 shows the number of scientific publications since 1995 that discuss exoplanets. Before 1995 practically nothing happened. As a result of the first discoveries a wide-spread interest took hold, and astronomy funds started to be diverted in this new direction. From about the year 2000 the results started to pour in. Since then the number of scientific reports has increased substantially, at least until 2009, and currently about 1000 reports per year are published. This in itself constitutes a revolution in astronomy and exoplanets have rapidly become one of the hottest research fields. In Fig. 6.6 we see the number of exoplanet discoveries per year. Discoveries made by different methods are marked with different colours. The trend is obviously increasing and there is little doubt that the explosive development will continue at an ever higher rate. Apart from the continued technological advancements, the time factor itself is important. In order to discover a planet at a longer distance from its host star you need considerable patience. Outlying planets may well take decades to complete just a single orbit around their host star.

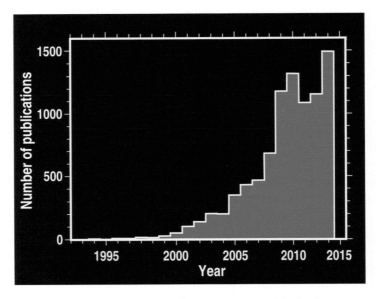

Fig. 6.5 Number of scientific exoplanet papers published per year (Peter Linde/ADS)

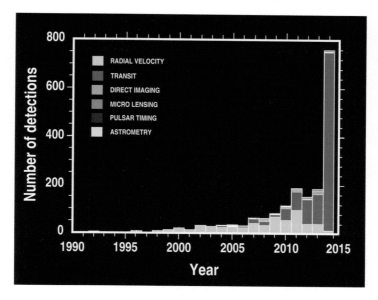

FIG. 6.6 Number of discovered exoplanets per year. The discovery methods are marked with different colours (Peter Linde/www.exoplanet.eu)

Currently there is a large number of observations of candidates where many are expected to be subsequently confirmed in the future. And new candidates are added all the time. From the figure it is also clear that discoveries made by the transit method are strongly increasing. The COROT and Kepler satellites are responsible for this as well as several ground-based projects previously discussed.

So what have we so far learned from all these discoveries of new planetary systems? We begin by taking a look at the properties of the stars that have planets.

Can All Stars Have Planets?

There is now sufficient statistical data in order for us to say something about the stars that today are known to have planets. This knowledge still is incomplete, but it is also true that stars in question are reasonably well characterised, much more so than their planets. We should also be acutely aware that much of the data suffers selection effects and thus does not necessarily represent a realistic sample. This being said, it is clear that the properties of

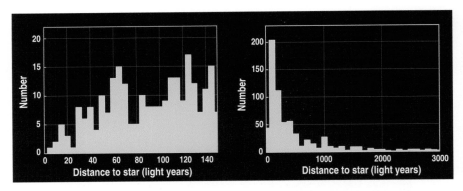

FIG. 6.7 The distribution of the distances to stars with known exoplanets. *Left*: Expanded part of the distribution with stars within 150 light years. *Right*: All included up to a distance of 3000 light years (Peter Linde/www.exoplanet.eu)

these stars can vary considerably, both in terms of size, mass and composition. Not surprisingly, there is no natural law prohibiting stars different from the Sun from having planets. In Fig. 6.7 we see, at left, the distance distribution for the stars now known to have planets. We find most of them within a distance of 150 light years—this part of the distribution is detailed in the diagram to the right. Both of the two major discovery techniques, the transit method and the radial velocity method, strongly favour stars that send a lot of light into the telescopes, and such stars tend to lie close to us. Also, transit detections need to be complemented by other types of observations in order to allow for a reasonably accurate distance estimate. For example, very few of the one thousand discoveries made by the Kepler satellite have any reliable distance estimates at all. Clearly there is good reason to believe that more distant stars may also have planets.

The mass of the star is its most important feature. To a large extent, mass determines its future evolution. Stars may have both smaller and larger masses than our Sun, a great majority have masses between 0.1 and 150 solar masses (see Box 6.4). The lighter stars are much more frequent than the heavy ones. Somewhat contrary to what one would expect, the low-mass stars have a dramatically longer lifetime. A star with a mass 25 times the Sun's has a stable lifetime of only about 6 million years while one having half the mass of the Sun is expected to survive for about 100 billion

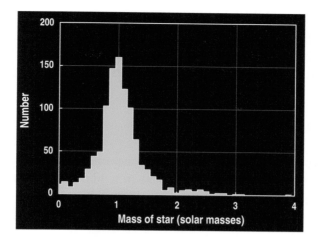

FIG. 6.8 The distribution of the masses of stars known to have exoplanets (Peter Linde/www.exoplanet.eu)

years. This can be compared to the expected life span for the Sun, which is about 10 billion years. This big difference is explained by the fact that the heaviest stars consume their available energy much more intensively than the low-mass stars. However, in all cases, when the nuclear fuel is near depletion all stars become unstable. They then expand to become giant stars, and in this process the outer layers may well be shredded back to space (see also Box 6.4). Does the mass of the star have any importance for the existence of planets? Probably not in a decisive way.

In Fig. 6.8 we see that the distribution of masses for the stars that have planets is rather wide. The record for heaviest star so far is about four solar masses. Again, the information here is incomplete. Spectra from heavier and therefore also much hotter stars have much fewer visible spectral lines, which means that the radial velocity method does not work very well. Additionally, such stars spin rapidly around their axes, making existing spectral lines broader and more difficult to measure accurately. This broadening of spectral lines is also due to the Doppler effect, since half of the star is receding (slight red shift) and the other half is moving towards us (slight blue shift). That exoplanets, in spite of this, have been detected around heavier stars is possible by observing such stars at the end of their evolution. As mentioned, they then expand to become red giants (see Fig. 6.4), which means both that the surface temperature goes down to about 3000 °C and that the rotation becomes slower. Subsequently, the radial velocity method can be applied. However, for red giants of any type, only planets far away will remain at a safe distance from their host star.

The Composition of Stars

An interesting question is whether the chemical composition of the star plays a role for the creation of a planetary system. A star mainly consists of hydrogen and helium, the two basic elements that were created when matter first formed, soon after the universe was created in a Big Bang, about 13.8 billion years ago. There is also a certain percentage of heavier elements in them; this fraction is usually referred to as the star's metal abundance. Astronomers have a simplified definition of metals. To them all elements heavier than helium are considered as metals. Heavier elements than hydrogen and helium exist only because they have successively been generated by nuclear reactions deep inside stars. When the first generations of stars become unstable they ejected into space matter enriched with newly produced heavier elements. After a (long) period of time, such matter can coalesce again and form new generations of stars. This becomes a cyclic process. Throughout billions of years, many generations of stars have been formed, with the result that later generations already from the beginning have been enriched with heavier elements. Our own Sun is an example of this. Out of the gas and dust disk from which a star is created, any potential planets will be formed. A planet of the Earth type is made of up to 90 % heavy elements like iron, oxygen, silicon and magnesium. We may ask ourselves whether the exoplanet data confirms a close relationship with high metallicity in the host star? Figure 6.9 illuminates this, based on data from about 1000 stars.

The measure of the metal content of a star is a little peculiar. It is designated [Fe/H] and specifies the amount of iron relative to hydrogen that is present in a star relative to our Sun. Iron is easy to identify in the spectrum of a star, and gives a representative, albeit incomplete, picture of the metal content of a star. The scale is logarithmic, with the value zero meaning a metal content equal to that of the Sun. −1 means a tenth of the Sun's, +1 means ten times

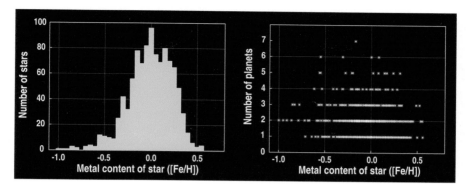

Fig. 6.9 *Left*: The distribution of the metal contents for stars known to have exoplanets. *Right*: Correlation between the metal content of the host stars with number of planets in their planetary systems (Peter Linde/www. exoplanet.eu)

higher, etc. The most metal poor stars known have a $[Fe/H] = -4$, which consequently means 1/10,000 of the Sun's metal content.

In Fig. 6.9 we see in the left diagram the current situation. A small majority of stars with planets have a similar or larger metal content than the Sun. This supports the idea that a star must be rather metal rich in order to have planets. But there clearly are exceptions: some stars with planets have a lower metal content than our Sun. As usual, the statistics are probably a little skewed, since most of the studies until now have concentrated on solar-type stars. But so far, no planet has been found around really metal poor stars. In the diagram to the right we see the relationship between metal content and number of detected planets. It appears that there is some preference for metal rich stars to have more planets.

The Distance, Orbital Period and Mass

Now let us take a look at the properties of the exoplanets themselves. Again, we should keep in mind that even though the number of discovered planets is rising all the time, many of the observations have a high degree of uncertainty. That could be true for the mass, the distance from host star or for any other assumed characteristics. Sometimes it is even unclear whether the existence of an exoplanet has been confirmed at all or whether the observation could

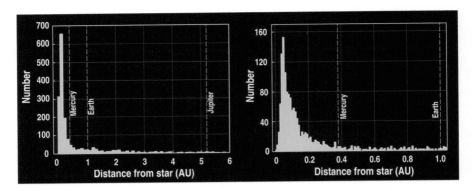

FIG. 6.10 The distance between exoplanets and their respective host star. *Left*: Exoplanets out to 6 AU. *Right*: The inner part of the diagram expanded to a maximum of 1 AU. The distances of Mercury, Earth and Jupiter are marked with *dashed lines* (Peter Linde/www.exoplanet.eu)

have some alternative explanation. The artistic illustrations that often accompany the discoveries (including some reproduced later in this chapter) are often quite bold and extrapolate the available information, to say the least. They should be taken with a grain of salt, but are nevertheless suggestive and imaginative.

Figure 6.10 illustrates the distribution of distance between known exoplanets and their host stars. The distances are given in astronomical units (AU), where 1 AU is defined as the Sun-Earth distance. The left diagram shows an overview out to 6 AU, the right shows the shorter distances in greater detail. For comparison, the solar distances for Mercury, Earth and Jupiter are marked with dashed lines. The diagrams show very few exoplanets with distances typical for the giants in our own system (Jupiter, Saturn, Uranus and Neptune), i.e. 5–30 AU. This is certainly due to incompleteness, which is not surprising, considering that the orbital period of, for example, Neptune is 167 years. It simply takes a long time to find such planets. The really remarkable feature of the diagram is rather that there is such an abundance of stars lying quite close to their host stars. More than half are closer to their star than Mercury, our innermost planet, is to the Sun! Even realising that there is a strong selection effect in the data, it came as a surprise that such planets would be so commonplace.

The orbital periods of the exoplanets around their stars follow the same pattern since distance and orbital period are closely

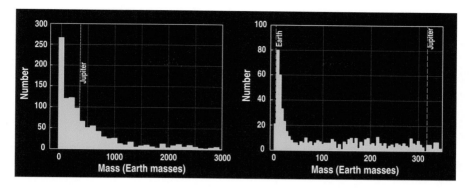

FIG. **6.11** The masses of the exoplanets. To the *left* is seen the complete sample, to the *right* is seen the lighter part of the distribution. The masses of Earth and Jupiter are marked with dashed lines (Peter Linde/www.exoplanet.eu)

connected via Kepler's third law. The third law tells us that the square of the orbital period (P) is directly proportional to the cube of the distance (a), i.e. $P^2 \sim a^3$. This means that almost half of all known exoplanets have orbital periods shorter than 50 days, to compared to Mercury's 88 days.

The mass of the planets is seen in the two diagrams of Fig. 6.11, measured in units of Earth masses. The masses of Jupiter and Earth are specially marked. To the right we see the inner part of the left diagram, magnified to illustrate the distribution of the lighter planets. As we can easily see, practically all known exoplanets are heavier than Earth. Again we have a very strong selection effect in favour of the heavier planets. Also, it should be noted that there is systematic underestimation for most of the measured masses. This is due to the fact that astronomers using the radial velocity-based measurements have to correct for a factor depending on the angle of the orbital plane of the stellar system in question with regard to Earth. That angle cannot be measured with this method, leading to an average underestimation of such mass determinations of about 15 %. In those cases where the same planet can also be observed with the transit method, this uncertainty is removed since then the orbital plane of the planet is known.

The diagrams show a clear trend, namely that planets with smaller masses are more common than the larger ones. Even if the number of small terrestrial planets is still insignificant, it is most

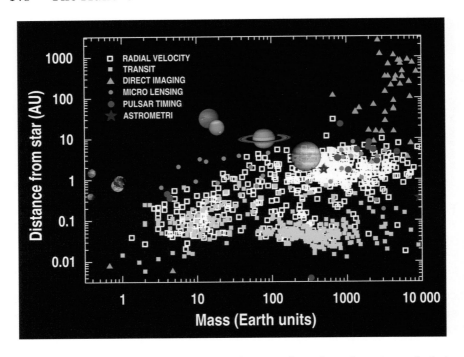

FIG. 6.12 The so far discovered exoplanets plotted as function of their masses and their distances to the host star. The different symbols denote the method used for discovery. For comparison, the planets of our solar system are marked. The inner rocky planets lie to the left of the diagram, while the outer gas giants are in the middle. Both axes are logarithmic and annotated in Earth units (Peter Linde/www.exoplanet.eu)

certainly explained by the great technological difficulty in finding them. There is every reason to believe that large quantities of small planets still await discovery.

In Fig. 6.12 we complicate things a little by connecting the two properties of mass and distance to the host star. Additionally, the different symbols show which method has been used for discovery. The scale on both axes is logarithmic in order to make it possible to review all information in a single diagram. This means that the change between successive tic marks is a factor of ten. The X-axis covers an interval between 0.3 and 10,000 Earth masses, while the Y-axis displays an interval between 0.03 and 3000 AU. For comparison, the planets of our own solar system are plotted in their respective positions in the diagram. As we can see, the four inner rocky planets end up far to the left, while the four giants are found approximately in the middle of the diagram.

A closer study of the figure reveals several interesting features, and provides a good illustration and summary of the current knowledge about exoplanet systems. First and foremost, we see that no Earth-type planets have yet been found. There are a few which have approximately the mass of the Earth but all of them are much closer to their sun than is the Earth, which implies that they should be very warm. However, once again, we have to remind ourselves that this part of the diagram most certainly will be filled in following increased sophistication of the observations. Another striking feature is the simple fact that our solar system does not seem to fit into the general picture of the current statistical data. Are we really so special, or is this also an effect of the incompleteness in the observations? We recall that planets at longer distances than 10 AU may take 30 years to complete one orbit which means that such discoveries will only be become possible to confirm with time.

Nevertheless it is interesting to notice that the data points occupy completely different parts of the diagram than does our own solar system. Evidently planetary systems can be formed in a variety of ways. At the same time, the points are not randomly located in the diagram. The data form three distinct clusters. The first lies at about 10 Earth masses and with a distance of about 0.1 AU. This mass is much larger than Earth's but is similar to that of Neptune. The distances to the host star are very short, as we see only 1/10 of the Earth's distance to the Sun. These planets are called hot Neptunes, and can best be described as melted ice giants. Another group lies at the same short distance but with larger mass (about 100–2000 Earth masses). We recognise these as hot Jupiters, which is the most straightforward discovery type for an exoplanet. Both these groups probably have their origin as gas giants and ice giants further out in their planetary systems, and have subsequently migrated inwards into orbits much closer to their host star. A third group has approximately the same masses but lies at "normal" distances, between 1 and 10 AU. Here we find the cold Jupiters, in many ways similar to the largest gas giant in our own solar system.

Box 6.5: Some New Exoplanet Terms

Planet type	Properties
Hot Jupiter	Hot gas planet with a mass larger than half a Jupiter mass. Lies close to its star, having a high atmospheric temperature
Hot Neptune	Molten ice giant with 3–20 % of Jupiter's mass. Lies close to its star
Cold Jupiter	Gas planet, similar to Jupiter. Lies at least 2 AU from its star
Super Jupiter	Gas planet with at least five times the mass of Jupiter
Nomad planet	Planet moving freely in interstellar space, not belonging to any star
Super Earth	Rocky planet with a mass 1.5–10 times the mass of Earth
Water planet	A super Earth with a large water mass fraction and a very deep global ocean
ExoEarth	A rocky planet, similar to Earth, with a mass less than 1.5 times the Earth's
Goldilocks planet	ExoEarth which lies in the habitable zone of its star
Twin Earth	Goldilocks planet with water and an oxygen atmosphere

From the figure we also see which detection methods have been most successful so far, along with what kind of results they preferentially give. Again, the radial velocity and transit methods dominate. It is clearly demonstrated how the transit method (filled squares) primarily works for planets lying close to their star. This is to be expected, since the probability is even lower that the orbital plane points directly towards the Earth to make a transit observable for exoplanets further away from their star. In addition, even if this is the case, it may well take years before a transit de facto takes place.

So far, discoveries based on direct imaging have only been possible for large planets at long distances from their host stars. These we find marked as triangles in the upper right corner in the figure.

As a result of the continuing exoplanet research we have gradually become exposed to some new denominations for exoplanets. Some we have already encountered. In Box 6.5 the new concepts are summarised. Let us take a closer look at a few of them.

Hot Jupiters and Neptunes

Astronomers have determined the masses of about one half of the exoplanets discovered so far. Two thirds of these have masses between a half and twenty Jupiter masses. Almost half of the planets

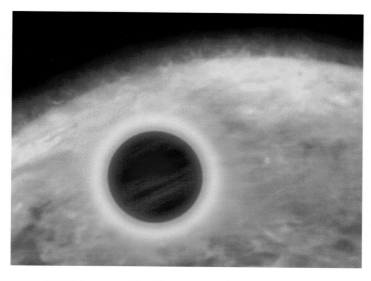

Fɪɢ. **6.13** WASP-33 b, a very hot Jupiter, with an estimated temperature of 3600 °C (NASA/ESA/G. Bacon (STSci))

in this sample lie in orbits closer to their star than does the Earth. This is the new category of planets with which we already have some familiarity. They are called hot Jupiters or—if their masses are a little smaller—hot Neptunes. There are a few quite extreme objects in this category.

The previously mentioned Hubble telescope survey of 180,000 distant stars unveiled a planet called SWEEPS-10. If the data is reliable, the planet orbits its star in only 10 h. It is so close to the host star that the temperature of its atmosphere is estimated to be about 2000 °C. Still hotter is apparently WASP-33 b (Fig. 6.13), where the temperature is estimated at an incredible 3600 °C. Considerably closer to us, just 150 light years away, we find HD 209458 b. This planet is also very close to its sun, and here the scientists have had the opportunity to study it in more detail. It turns out that water vapour apparently is leaving the planet and seems to leave a streak behind the moving planet (Fig. 6.14).

Several of the hot Jupiters also have have very low densities—in some cases just a fifth of our Jupiter's. This is interpreted as swelling, caused by the high temperatures.

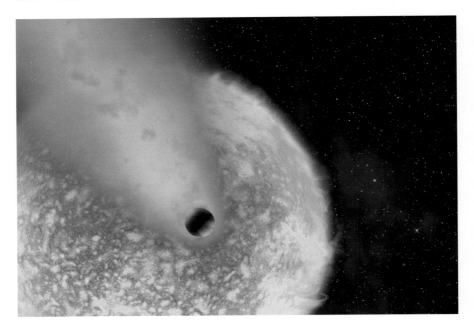

Fig. 6.14 The planet HD 209458 b, another hot Jupiter, shows signs of successively losing water vapour into space (ESA/Alfred Vidal-Madjar (IAP, CNRS)/NASA)

Super Earths

A new and exciting category of planets is the super Earths. About 10 % of the exoplanets with measured masses belong to this category. We have no comparable planets in our solar system, but the super Earths are probably quite common in others. The interesting thing about super Earths is that among them there should be many rocky planets offering better conditions for life than many other planets. As is evident from the name, super Earths are larger and heavier than the Earth, up to about ten Earth masses. However, it should be emphasised that the term super Earth does not imply Earth-like conditions, usually very little is known about these planets, except their masses. Figure 6.15 shows a scenario which could be found on the planet Gliese 667 Cc, a super Earth orbiting a star which in turn is a member of a triple star system.

FIG. **6.15** A conceivable view from the planet Gliese 667 Cc, a super Earth orbiting the star Gliese 667C, which in turn is part of a triple star system (ESO/L. Calçada)

Multiple Planet Systems

There are currently close to 500 systems where more than one planet has been discovered. More than 150 of them have more than two planets. Again, more complete information will certainly be available in time. That several planets exist in the same system is hardly surprising, given what we now know about planetary system formation. In 1999, astronomers documented the first multi-planet system, Upsilon Andromedae. To date, four have been detected, however, all of them seem to be of the Jupiter type. Planet d seems to lie at a reasonable distance from the host star, it has been speculated that a moon orbiting such a planet could be a suitable place for life (also compare Chap. 4). An artist's conception of a possible scenario is seen in Fig. 6.16. Kepler-11 is one of the more "complete" systems where six planets have been detected

FIG. 6.16 An illustration of a possible view from a fictitious moon orbiting υ Andromedae d, a giant planet in the first known multiple planetary system (Luciano Mendez/Wikipedia)

so far, all of them in orbits close to their star, which by the way is similar to our own Sun. The Kepler-11 system seems to be an example of a more "normal" planetary system in the sense that it actually shows similarities to our own (Fig. 6.17). All these planets have been detected using the transit method; there have even been several transits observed at the same time. Their masses are larger than Earth's; while at the same time their densities are lower. At least three of the planets can be characterised as hot Neptunes; hardly any seem to harbour terrestrial conditions.

The star HD 10180 is also considered to have at least six planets. Here the observations have been made with the radial velocity method. Deriving the existence of so many planets requires both sophisticated observations and analysis. Simply pondering the movements of the Sun in Fig. 5.2, Chap. 5, illuminates how complicated and minimal the motions of the stars are in this type of research. We also have a similar situation with Kepler-11, where the planets have larger masses than Earth and are located much closer to their star. In Fig. 6.18, examples of a few multiple planet systems are shown.

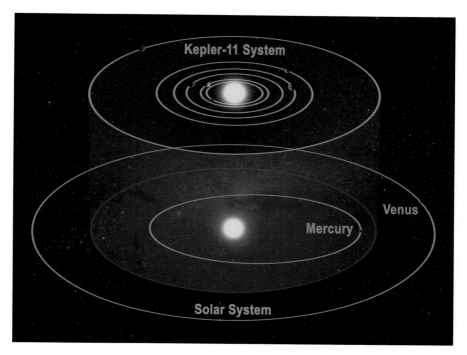

FIG. 6.17 A number of multiple planet systems are known. This is a comparison between the Kepler-11 system and our own solar system (NASA/T. Pyle)

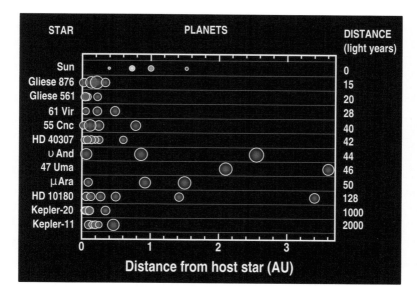

FIG. 6.18 Example of a few planetary systems with several planets. At the *top* our own solar system is shown for comparison. To the *right* are given the distances to the respective systems (Peter Linde/www.exoplanet.eu)

Small Planets

In 2011 the Kepler satellite science team for the first time reported
time that a exoplanet smaller than Earth had been detected. The
discovery involves the star Kepler-20, at a distance of about 1000
light years and very similar to the Sun. The team detected two further planets, in addition to the three previously known. The first
three are considered to be warm Neptune type planets, but the two
latest discoveries have sizes similar to the Earth's. All five planets
lie very close to their star and none of them have conditions suitable for life. The assumed temperatures for the two small planets
are estimated at 400 °C for one and about 700 °C for the other.

In the beginning of 2012, the discovery of three small planets
around the red dwarf star Kepler-42 were announced (see Fig. 6.19).
All of them are smaller than Earth, the smallest about the size of
Mars. The star is about five times smaller than our Sun, and this
facilitates finding small planets when using the transit method.
This is due to the fact that even a small planet covers a larger
fraction of the light of a small star than it would for a larger star.
These discoveries demonstrated Kepler's capability to detect planets

Fig. 6.19 A possible view from near Kepler-42 d, the so far smallest known
planet (NASA/JPL/Caltech)

the size of Earth, and continued studies have also resulted in a few such discoveries of planets inside the habitable zone.

As of Nov 2015, Kepler has reported almost 4700 suspected but unverified cases of exoplanets. Several groups of scientists are now looking closer at this sample, observing and analysing the data in different ways. There are good reasons to believe that the majority (perhaps as many as 80 %) will turn out to be planets, once it has become possible to examine the observations more closely. Results so far indicate that smaller planets of the super Earth or Earth type can be formed also around more metal poor stars than the Sun. On the other hand, it seems that gas giants of the Jupiter type are preferentially formed around more metal-rich stars. The conclusion is anyway that smaller, Earth-like, planets should be more common than the gas giants, many of which have already been discovered.

Water Worlds

In the menagerie of new planets, researchers speculate about water worlds, also known as ocean planets. Such planets would originate far out in a solar system, outside the snow line, and would initially consist mainly of ice.

The current view is that such planets may migrate closer to their suns, gradually passing the snow line and then largely melting into liquid water. The whole planet would then be covered in water, perhaps with depths of hundreds or even thousands of kilometres. In 2003 a group of French astronomers published a model of what a typical planet could look like. Figure 6.20 shows a cross-section of such a hypothetical water planet compared to Earth. In this example, the model mass was set to six Earth masses, in between the masses of Earth and Neptune. The recipe is (1) Earth mass metals, (2) Earth masses minerals and (3) Earth masses ice. In this case, the ocean depth becomes about 5000 km. Above it there is a 100 km thick atmosphere. Interesting physical phenomena would occur both at the bottom of the ocean and on the surface. The pressure of thousands of kilometres of water would create new forms of ice at the bottom. At the other extreme, if the planet was sufficiently close to its sun, the ocean would be boiling and create a very dense water vapour-based atmosphere.

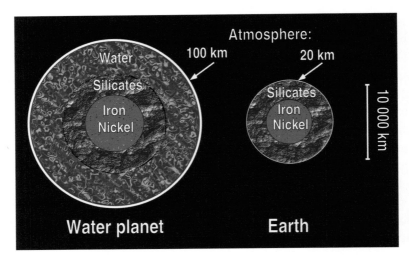

FIG. 6.20 Simplified model of a water planet, compared to the composition of the Earth. The respective atmospheric heights are given (Peter Linde)

Increasingly strong evidence continues to accrue for the existence of such planets. One of the most interesting candidates is GJ 1214 b (see Fig. 6.21), which was discovered in 2009. The planet is a super Earth with a mass six times that of the Earth. It is orbiting a red dwarf at close range. The density of the planet is twice that of water but just half that of the Earth's. This tells us something of its composition. It is evidently not a purely rocky planet, as the fraction of heavy minerals and metals must be less than on Earth. Instead, a large component could be water. During 2012 the Hubble telescope observed it again, and further conclusions about its atmosphere could be drawn, characterising it as water vapour.

Speculations about life existing and developing in such environments are not very far-fetched and are frequent in the science fiction literature. We will return to this question later, but will nevertheless emphasise that it is an intriguing subject. Could a civilisation develop in a 1000 km deep ocean with no land anywhere? Clearly, even a modest depth would provide strong protection from dangerous radiation and climatic variations at the surface.

In our solar system there is no such planet, but it is known that several of the larger moons of the giant planets consist of large components of ice. This is probably true also for the inner parts of Uranus and Neptune, even though they are surrounded by a dense

Fɪɢ. **6.21** The planet GJ 1214 b lies at a red dwarf star. The planet may to a large extent consist of water (NASA/ESA/D. Aguilar/Harvard-Smithsonian Center for Astrophysics)

atmosphere of hydrogen and helium. In a distant future, when the Sun expands to become a red giant star, it is conceivable that these celestial bodies will be more suitable to life than they are now.

Bound Rotation and Strange Orbits

Many of the exoplanets have orbital properties that differ dramatically from what we see in our solar system. We have already noted that some are located very close to their host star. This probably leads to an unavoidable side effect; such planets in most cases will have bound rotation. For an exoplanet in immediate proximity to its star this means that one side will have intense irradiation while the other is in darkness. Whether eternal heat and cold rule in the two respective hemispheres depends on the existence of an atmosphere.

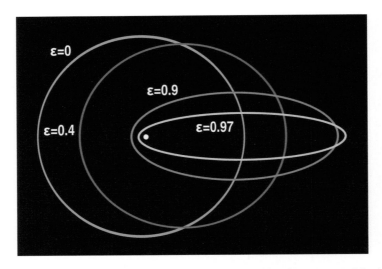

Fig. 6.22 Examples of some eccentric orbits. The largest width (major axis) is the same for each value of the eccentricity. The host star lies in one of the two foci. For comparison, the eccentricity of the Earth's orbit is 0.017 (Peter Linde)

If one exists and is sufficiently dense, it is conceivable that wind systems and heat conduction distribute the energy reasonably well over the whole planet.

Many of the orbits of the exoplanets are rather oblong. The deviation from a circular form is called eccentricity, and is expressed with the parameter ε (epsilon), where the value 0 represents a perfect circle. Figure 6.22 shows examples of different elliptical orbits. Half the major axis, i.e. half of the greatest width of the ellipse, is the same for all orbits in the figure. A planet which moves in an elliptical orbit has, according to the first law of Kepler, its star in one of the two foci of the ellipse. This means that the shortest and longest distances will vary significantly in orbits with high eccentricity.

All planets in our solar system have reasonably round orbits with an eccentricity less than 0.1 (except for Mercury). This is also true for about half of the known exoplanets. But almost 20 % have eccentricities larger than 0.4, with the current record at 0.97. At an eccentricity 0.4, a planet receives five times as much heat from its star at its minimum distance as it does at its maximum. This is not a favourable condition for the emergence of life.

It is well known that in our solar system all the planets lie practically in the same plane, known as the ecliptic. The rotational axis of the Sun is perpendicular to that plane. This agrees with the generally accepted view of how a planetary system is formed from a rotating disk. But what about the orbital planes of the exoplanets and their relation to the rotational axis of their respective host star? Here, the information is still scarce since it is quite difficult to obtain reliable and detailed measurements. But there are indications that quite strongly inclined orbital planes are not uncommon. There are also a couple of cases where there are reasons to believe that the planet circles in the opposite (retrograde) direction relative the rotation of the star. We will return to possible explanations for this, but it is evident that the creation of a planetary system can cause many dramatic and chaotic developments.

Nomad Planets

The possibility of finding planets roaming empty space without belonging to any star is perhaps not inconceivable. But the possibility of their discovery is certainly quite remarkable. It is easily realised that transit or radial velocity measurements are out of the question since they in various ways fully depend on observing the light from the host star. But the third main method, microlensing, works in this case. In May 2011 a Japanese and New Zealander research group reported that after studying more than 50 million stars they had observed a number of very short-lived microlensing phenomena, where the change in brightness lasted less than two days. This is interpreted as caused by bodies of the Jupiter class. No accompanying stars could be identified, even with the support of a large 8 m telescope in Chile. The teams assert that the data implies free-floating planets, also known as nomad planets or rogue planets (see Fig. 6.23). The conditions on such a planet ought to be rather harsh, just the contrary to what you would expect on, for example, a hot Jupiter. Without a host star, the only available light would come from the stellar sky, and it is very cold. Some heat may remain after the planet's formation and there might be some contribution from radioactive decay, as in the case of the Earth. But a probable place for life it is not, unless you consider

Fig. 6.23 A hypothetical free-flying nomad planet, approximately the size of Jupiter, is moving in eternal darkness and bitter cold (NASA/JPL-Caltech)

the amazing scenario in Arthur C Clarkes brilliant short story "Crusade" from 1966, where he speculates on a global intelligence based on currents in superconducting liquids…

The origin of the nomad planets is not well understood, but most scientists are inclined to believe that they were hurled away during star and planetary system creation in stellar clusters. Conditions then were very chaotic and, when bodies came too close to each other, orbits may have been drastically altered via strong gravitational pulls, resulting in both stars and planets simply being expelled from their previous environment.

Unusual Planetary Systems

An established fact today is that almost half of all stars are created as couples of double stars orbiting each other. Some have orbital periods of hours, others of hundreds of years. There are also even more complicated systems with three or four stars that circle around each other in different ways. For double stars that are separated by considerable distances, perhaps 1000 AU, and where one of the components is a small red dwarf, it is not so strange to believe that the larger star could harbour a planetary system. Several such cases are already known. But for tight double stars the situation is different. Scientists did not have any expectations of finding planets among such chaotic conditions, but yet again nature has surprised us. A handful of such cases have recently been observed, and more or less verified. On of the confirmed cases is Kepler-16 b, discovered in 2011. It is a gas giant, somewhat smaller than Jupiter, that actually orbits a double star.

So again, science fiction has become reality. In the famous movie series Star Wars, the planet Tatooine plays an important role since it is the planet on which hero Luke Skywalker was brought up. In the film it can be noted that this planet has two suns in the sky. The same applies to Kepler-16 b. The two stars are both smaller than the Sun, one is orange and the other a red dwarf star. They orbit each other once every 41 days. The planet, in turn, moves around both of the stars in a nearly circular orbit in 229 days, at approximately the same distance as Venus in our solar system. However, it is not as warm as Venus; on the contrary it is a cold gas giant which is outside the habitable zone. This is explained by the fact that the two stars, even together, do not emit as much heat as does our Sun. A (very hypothetical) inhabitant there would in any case have a quite spectacular sky with several suns (Figs. 6.24 and 6.25).

The Atmospheres of the Exoplanets

It is impressive to follow the discoveries of exoplanets and how they are arriving in rapid succession, and continuing to do so. Orbital properties and masses have been determined for about 1000 exoplanets. But getting information about the atmospheres of the exoplanets and their composition was considered highly improbable as late as the turn of the millennium. Nevertheless, the ingenuity, creativity and diligence of scientists has already led to rough analysis of more than two dozens exoplanet atmospheres. A number of different molecules have been detected, there have been studies of temperature differences between day and night

Fig. 6.24 View from the planet Kepler 16 b. Star War hero Luke Skywalker grew up on Tatooine, which similar to Kepler 16 b, also had two suns (NASA/JPL-Caltech/T. Pyle)

Fig. 6.25 An imaginative view from a hypothetical moon belonging to a giant planet orbiting the triple star system HD 188753. Some scientists though question if this planet even exists (NASA/JPL-Caltech)

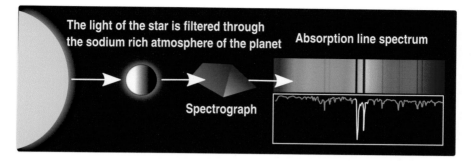

FIG. **6.26** The principle for detecting sodium in the atmosphere of the planet HD 209458 b. The atmosphere of the planet affects the spectrum of the light from the star (Peter Linde)

sides of the planet, and even of the conditions at different altitudes within the atmospheres.

As we have already discussed, it is possible to extract information during an exoplanet transit by special analysis of the tiny fraction of light that passes through the atmosphere of the exoplanet. Since all observations are digital you can simply compare the spectrum of the light from the star during and after (or before) the transit. The result is an extremely weak but nevertheless purified version of the atmospheric spectrum of the exoplanet. In Fig. 6.26 we see the principle for how such a detection can be made. The problem is somewhat simplified by the rich abundance of hot Jupiters. In Fig. 6.3 we see that a hot Jupiter is perhaps "only" a thousand times fainter than its host star, at least in the infrared part of the spectrum. Through its proximity to the host star it becomes strongly illuminated, and therefore brighter and easier to detect. In addition, the high temperature in its atmosphere, perhaps 1500–2000 °C, facilitates finding spectral features in the light. An example of such a planet is HD 189733 b (see Fig. 6.27).

The Closest Exoplanets

How close are we to finding a planet suitable for life in our neighbourhood? Well, this goal has certainly not yet been reached, but we are well under way. Within a radius of 50 light years, 91 exoplanets have been found so far, in 46 different planetary systems.

FIG. 6.27 HD 189733 b is a hot Jupiter (900 °C) orbiting so close to its star that the orbital period only is 2 days. In its atmosphere water vapour, methane and organic molecules have been found (ESA/C. Carreau)

Figure 6.28 plots them using their host star's distance to us as one variable (X-axis) and their distance to their respective host star as the other (Y-axis). In the figure different colours are used to represent the number planets in each system. Since every planet in a certain planetary system has approximately the same distance to us, it is possible to read off in the vertical direction which ones belong together. A well-known example, the five-planet system 55 Cnc, is marked with white dots in the diagram. In 2015, a report claimed to have observed strong temperature variations on 55 Cnc e, possibly explained by heavy and variable volcanic activity (see Fig. 6.30).

Approximate masses are also indicated by the internal sizes of the dots. The largest are larger than Jupiter, the smallest are super Earths. One nearby exoplanet is Epsilon Eridani b, at a distance of 10 light years. It is a cold Jupiter-like planet in an eccentric

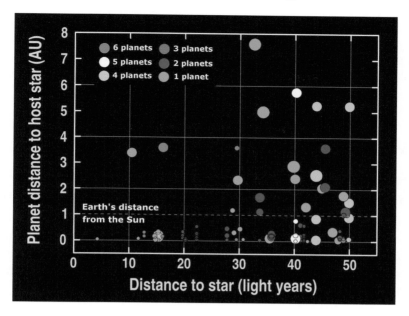

FIG. 6.28 Exoplanets for stars in the proximity of the Sun. The sizes of the symbols are proportional to the mass of the planet. The *colours* represent number of planets in systems with several discovered planets (Peter Linde/www.exoplanet.eu)

orbit far away from the habitable zone. About twenty qualify as super Earths, but as can be seen from the diagram they are all located close to their host star. The closest we come to a planet similar to Earth seems to be Gliese 581d. It is six times heavier than Earth and lies at moderate distance from the star, which is a red dwarf and considerably smaller than the Sun. A closer examination reveals that it lies at least partly in the habitable zone of the star. But its orbit is strongly eccentric, and since the size of the planet is not known it is not clear whether it is a rocky planet.

In 2012 the first discovery of an exoplanet in the α Centauri system was announced. α Centauri consists of a triple star system which, at a distance of 4.3 light years, is our closest neighbour. After 4 years of industrious observations with the HARPS spectrograph connected to the ESO 3.6 m telescope in Chile, scientists finally determined that α Centauri showed a very small but measurable Doppler effect caused by the gravitation of an exoplanet.

The orbital period of the planet is short, just 3.2 days, which means that it lies very close to its star and must be very hot. But it is not a hot Jupiter. Rather, it has a mass comparable to the mass of the Earth. The discovery is very exciting, as other planets may well exist in the system that offer more life-friendly environments.

The Habitable Zone

In Chap. 4 we discussed the meaning of the habitable zone notion. We found that it was a flexible concept which depended on a variety of different conditions. In a wider context habitable zones can exist around practically all stars that are reasonably stable. Naturally, this zone attracts special attention in the exoplanet research. So, what is the status among the almost 2000 known exoplanets? Are there any in this interesting region?

During 2011 NASA presented such a discovery among the observations made by the Kepler satellite. Kepler-22 b, at a distance of about 600 light years, seems to satisfy the requirements. In Fig. 6.29 we see a comparison between the habitable zones

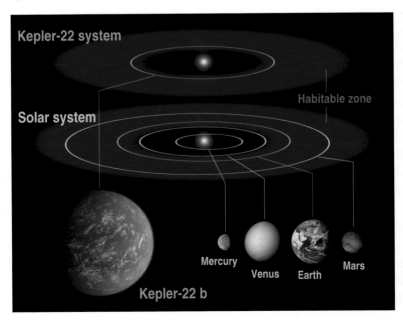

FIG. 6.29 Kepler-22 b, a super Earth which probably lies in the habitable zone of its host star (NASA)

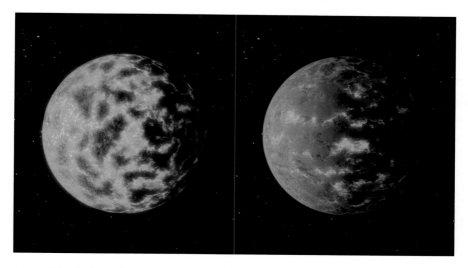

FIG. 6.30 55 Cnc d is an exoplanet that has been shown to exhibit strong temperature variations, possibly explained by a strong and variable volcanism (NASA/JPL–Caltech/R. Hurt)

for Kepler-22 and our own solar system. The star in question is quite similar to our Sun, and the planet may possibly be classified as a super Earth, but very little is known about it. A more recent discovery was announced in 2015. Kepler-452 b is a planet at the right temperature within the habitable zone, and only about one-and-a-half times the diameter of Earth, circling a star similar to our own Sun, but about a billion years older. As for many of the Kepler discoveries, complimentary mass determinations are necessary in order to establish the nature of the planet. This allows for determination of density, which makes it possible to distinguish between gaseous and rocky planets.

There are a few more planets that are located more or less at a suitable distance from their host star. On the internet, special sites are dedicated to this kind of data. The Habitable Exoplanets Catalogue, which is maintained by the university of Puerto Rico, currently counts about ten cases of planets with masses similar to the mass of the Earth, which are supposed to lie inside the habitable zones of their host stars. The Habitable Zone Gallery is another website which visually illustrates where the known planets are situated relative to the respective habitable zones.

Improved Theories

Twenty years after the first discoveries, it has become apparent that the conventional view of what a planetary system should look like, and how it is formed, to say the least has received several dents. Currently it looks more like our own solar system is the one that stands out as rather unusual. We have already pointed out several important phenomena that do not have any equivalence in our own solar system. Hot Jupiters, elliptical orbits, super Earths and water planets are the most striking examples. Naturally, this new wealth of information means that the theories for planetary system formation need to be complemented and improved. For example, how do we explain the hot Jupiters?

One of the main theories is that these giant planets, like our own, once were created far from their host star. We remind ourselves from basic theory (Chap. 2) that planets are created in a rotating disk of gas and dust. Smaller rocky planets are preferentially formed inside the snow line, gas giants outside. But due to the properties of the host star (among other things), the gas in the disk is, with time, pushed away and disappears into space. Today, our own solar system is completely empty of that gas which once must have existed and accumulated into the gas giants. So, one interesting point is the question of when the gas appears in the formation process, before or after the creation of the gas giants. Current beliefs among some scientists are that the hot Jupiters in some systems were more or less finished with their formation phase before the disappearance of the gas. Thus they were simply exposed to a braking force which gradually led them to migrate inwards in the system, finally ending up close to their host star.

The many elliptical orbits also need explanation. The details of this remain for the scientists to account for, but it is clear that the chaotic conditions and gravitational interactions prevailing during the formation of a planetary system led to very elliptical orbits, and sometimes in very oblique angles relative to the expected orbital plane.

Consequently, astronomers now have many more facts to include in their theories for the formation of planetary systems. Great advances in the understanding of this process have been made, but many things still remain to be untangled.

So How Many Planets Are There?

Until 1781, six planets were known. Then William Herschel discovered Uranus. The discovery of Neptune followed in 1846, and that of Pluto in 1930. At that point nine planets were known. This was later decreased to eight again when Pluto was degraded to dwarf planet. Currently, almost 2000 planets have been discovered and several thousands more are suspected cases. So, how many planets are there in our own galaxy, the Milky Way? A definite answer certainly is not available; we are only at the beginning of an exciting journey. Estimations of the number of exoplanets can only be based on statistical data. Nevertheless, some estimations are quite reliable, especially concerning those types of planets which are easy to detect and where statistical data has become rather complete. For example, if we limit ourselves to hot Jupiters at solar-type stars, observations show that such planets exist at about 1–2 % at these stars. In a more general sense the science group of Michel Mayor could, in 2011, report that based on radial velocity measurements, at least 50 % of all solar-type stars had at least one planet. This still did not include those with a longer orbital period than 100 days, which ought to be quite a few. Added to this general picture should be the fact that planets with small masses are more common, and those statistics are very sketchy.

A 2013 report containing further analysis of Kepler data indicates that about 22 % of the stars have Neptune-sized planets, 20 % have super Earths and around 16 % have Earth-sized planets. Another statistical estimate comes from micro lensing observations. In the beginning of 2012 a large international group of scientists published a report based on several years of such observations. The method is more or less independent of type of star. They had been looking for planets down to five Earth masses and at a distance between 0.5 and 10 AU. This is a fairly representative target group, with the exception that the smaller terrestrial planets still are missing. In spite of having observed just a handful of systems, they boldly concluded that planets are very commonplace. On the average, every star would have 1–2 planets, suggesting a galaxy containing far more than 100 billion planets.

From this it should be evident that exoplanet research has a bright and exciting future.

7. Exoplanets: A Look Into the Future

What is the most fundamental driving force in the ongoing investigation of the exoplanets? It is not far-fetched to assume that finding life on another planet is at the top of the list. A discovery like that would be revolutionising and thus constitute a strong motivation for the scientists. The research within the field also attracts the interest of the public, which in turn improves the possibilities of the researchers to fund new observational projects. What is realistic to achieve during the next few decades?

Assume that we would like to find an exoplanet out there similar to Earth, within a reasonable distance. We would also like to determine whether there are traces of life on it. How would we go about doing it? We begin by deciding on a reasonable distance and choose 10 parsecs, a unit of length that astronomers often prefer to use. One parsec equals 3.26 light years. Within the distance of 10 parsecs there are about 70 stars which are rather similar to our Sun. By "rather similar" we mean that they have a mass which differs less than 20 % from that of the Sun's. Some of the 70 stars are unstable and vary in their luminosity, but 45 of them seem to be suitable candidates for a closer inspection for terrestrial planets.

For such planets, astronomers want to be able to measure the orbit, the mass, the spectrum in visible and in infra-red light, and to assess how these properties change with time. If such measurements are successful, it becomes possible to identify a number of interesting properties, such as temperature, size, mass, density, gravity and reflectivity. Tools that are being further optimised for this purpose include transmission spectra, secondary eclipse curves and high resolution spectra. A first step will be to improve simultaneous measurements of exoplanet radii and masses. This will determine the density of the planet, enabling the important distinction between rocky super Earths and small gas planets.

© Springer International Publishing Switzerland 2016
P. Linde, *The Hunt for Alien Life*, Astronomers' Universe,
DOI 10.1007/978-3-319-24118-0_7

If, by happy coincidence, an Earth twin is discovered among these measurements, further investigation will focus on identifying additional life-sustaining conditions. These could include atmospheric pressure and the presence of clouds, presence of liquid water on the surface, possibilities for plate tectonics and the nature of the surface. Even more detailed studies could reveal the length of the day (i.e. rotation time of the planet), existence of weather systems and meteorological seasons. After that would follow a direct search for bioindicators.

Bioindicators

When searching for life on other planets, which are the main factors that show evidence of life? Which indicators for life do we have the capacity to look for?

In our own solar system we can conclude that observing from a distance does not reveal any signs of life outside the Earth. Unfortunately (or perhaps fortunately?) the numerous Martian canals observed by Percival Lowell were only illusory. An invasion from there is not to be expected. The possibility of finding life in our own solar system is, however, not exhausted, at least for primitive life forms or remnants from such. But to be able to establish the truth on this we probably need to investigate the matter at close range (see Chap. 4). However, to visit planets in other star systems will not become possible until far into the future (see Chap. 12). Are there, instead, any exoplanets that can be observed from Earth in sufficient detail to establish the existence of life?

The exoplanet research can be divided into three phases. The first, which deals with proving the existence of exoplanets, is completed. The second phase, to successively discover more exoplanets, is now well under way, with major efforts in progress at the moment. The third phase, to characterise the exoplanets, has just begun. As we have seen, the largest telescopes in space and on the ground are being used for this purpose, and there are plans for new advanced space probes specialising in follow-up studies of the newly discovered exoplanets.

If, for a moment, we disregard the remote possibility that a civilisation on an exoplanet reveals itself through signals of various kinds (see Chap. 11), then the more realistic question is whether we can detect primitive life forms from a distance. The answer is yes, although with enormous difficulties. What we can study is the extremely faint light that an exoplanet reflects from its parent star.

As usual, let us use our own solar system as our starting point. How easy is it to actually discover the Earth as seen from the stars? Two factors are decisive: firstly, how bright the Earth shines and secondly, how far from the Sun it lies as seen from the distance of the observer. Both factors are major problems. For example, to detect the Earth from the planet Jupiter is simple. Photos of this kind taken by interplanetary spacecraft are already available. But, as we know, the distances to the stars are enormously much greater.

As an example, let as imagine that we are looking at the Sun and the Earth from a hypothetical planet orbiting a star at a distance of 20 light years. The Sun is an easily visible star in this planet's night sky, albeit not very bright. But the Earth would be about 1.5 billion times fainter (see Fig. 6.3)! The Earth's magnitude, a measure the astronomers use to define brightness, would be about 27, which is very faint indeed. Visual photons from such a light source arrives at Earth less than once per second per square metre! But this is not the most serious problem. Such observations are already within reach of today's largest telescopes, not to mention tomorrow's. The biggest problem is that the Earth, seen from this distance, lies very close to the Sun, approximately 1/6 of a second of arc. This small angle can be compared to the distance between the two headlights of a car, as seen from a distance of 3000 km. To make matters worse, one of the headlights must be reduced to a small firefly!

Discovering this Earth-like planet is, however, not enough. We also want to establish if there are any life signs. Within the near future, advances will reach even this level of accuracy. Of course, other problems always arise, the problem of defining what life really is, and the possibility of alternative life forms. But to carry out any type of scientific analysis we have to proceed from the facts that are available, which means that we are looking for life similar to what we know on Earth.

What is needed to do is so-called spectroscopic analysis of the light from the planet. This means that you study light of different wavelengths, i.e. the spectrum of the planet. Since the planet only shines by reflected light from its parent star, we may perhaps expect to see only the spectrum of the parent star, albeit indirectly. But, as we have seen in the previous chapter, the planet actually affects the light it reflects. And here is the chance to detect signs of life.

Again, we take our earthly experiences as a starting point. In 1993, the well-known American astronomer Carl Sagan published an article outlining how signs of life would be observable from the Earth's reflected light. The publication was based on observations made by the spacecraft Galileo, which was on its way to Jupiter. In the beginning of its voyage it passed the Earth at a distance of a couple of thousands kilometres, in order to further increase its speed. To study the Earth, Sagan used several of the instruments on board the spacecraft. The result was that conclusive evidence for life on Earth was found, even intelligent life! Spectroscopic results distinguished traces of elements such as water, oxygen, ozone, carbon dioxide, carbon monoxide, nitrous oxide and methane. The images from the on-board cameras showed large amounts of liquid water but the limited resolution available prohibited any detection of artificial construction. On the other hand, radio signals were detected from the dark hemisphere which clearly contained information and thus indicated intelligence.

Recently, scientists have tested another method, more suitable for application to exoplanets. One of the world's largest telescopes, the European VLT (Very Large Telescope), observed our closest neighbour in space, the Moon. Seen from the Moon, the Sun is below the horizon during the lunar night, but the Earth can be positioned high in the lunar sky. It is also well known that a full moon illuminates quite a lot during a dark night on Earth, but on the Moon a "full Earth" shines considerably brighter since the Earth is four times larger. In turn, this light from the Earth is reflected back to Earth and can often be seen by naked eye as a greyish faint part of the lunar disc. It was this light that was observed by the VLT. The scientists carefully analysed both the colour and the polarisation of light from Earth after it had been reflected off the lunar surface, just as if it would have come from an exoplanet. They could show that the Earth's atmosphere partly contains clouds, that part of the Earth is covered with water and—most important of all—that there is vegetation. They could even see how the cloud cover changed and how the amount of vegetation changed when different parts of the Earth reflected light towards the Moon.

To find interesting information in the Earth's light at close range is evidently possible, but to find it in the faint light from a planet close to its parent star is a major technical challenge.

Fortunately, as regards tracing evidence of life on exoplanets, the situation is improved by the fact that some of the best so-called spectroscopic biomarkers can be observed in the infra-red part of the spectrum (see Fig. 6.3). There, the light of the Sun and other stars is less blinding and the Earth is "only" about two million times fainter.

The elements which primarily are associated with the presence of life are water vapour, oxygen, ozone and methane. It should be possible to detect these in the light from an exoplanet. The presence of water vapour suppresses the whole spectrum to some extent, and may suggest large quantities of liquid water. The others can be recognised via so-called absorption lines in the spectrum (Fig. 7.1). Ozone is easy to observe even in small quantities and its presence implies a high oxygen concentration, since the ozone is continuously created from ultraviolet light hitting oxygen molecules. Methane gas is produced by microbes; a

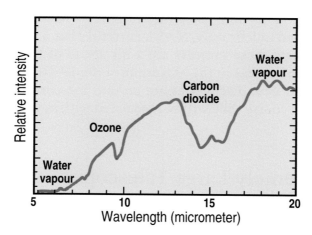

FIG. 7.1 The infrared part of a simulated spectrum of the reflected light from a terrestrial exoplanet. Some absorption lines interpreted as biomarkers are indicated (Peter Linde)

well-known example is digestion processes from cows and other herbivores. Lately, some scientists are suggesting that industrial pollution elements may also be possible to identify in the light from exoplanets.

Success—But Also Concerns

The technological and intellectual resources made available for exoplanet research have, during the last fifteen years, borne fruit in a way that nobody could imagine in 1995. Dedicated astronomers have used both small and large telescopes in the most ingenious ways. Sophisticated software has played a major role in automating fundamental surveys. The theoreticians have had a field day, using a stream of new data to develop models for phenomena that hardly existed in the real world until very recently. Yet, there is no doubt that the progress of the field is still in its infancy. So what does the near future look like?

There are clouds in the exoplanet sky. Unfortunately, several vaunted projects have been tossed in the waste-paper basket, or at least shelved. The European Darwin project was a gigantic project which envisioned several Earth-orbiting satellites interacting to form one single telescope to directly observe even small and terrestrial exoplanets. The similar American project Terrestrial Planet Finder (TPF) and Space Interferometry Mission (SIM) have been postponed indefinitely. NASA currently has a very restrained budget. Perhaps these projects are a bit ahead of their time, like the Kepler project was in the beginning of the 1990s. Nevertheless, it is clear that new discoveries are around the corner in the near future. And of course they will be associated with new technological advances.

The Extremely Large Telescopes

Today's largest ground-based telescopes have monolithic one piece mirrors with diameters of eight metres. To precision grind such beasts with an accuracy of a few nanometres is, in itself, a spectacular technological achievement. But designers have reached the current size limit. Instead, the latest efforts focus on constructing telescopes with segmented mirrors. A number of smaller mirror elements are grounded and subsequently put together with great precision to a single large mirror surface. The technology is already being used and has been shown to be successful. So far, the largest telescopes of this type have 36 segments, making up a single 10 m diameter mirror. Examples are the Keck telescopes in Hawaii (see Fig. 5.17) and the Gran Telescopio Canarias (GTC) on La Palma, one of the Canary Islands. But this is only the beginning. The next generation of extremely large telescopes is under way.

Three such telescopes are currently under construction. Two are primarily American, The Thirty Meter Telescope (TMT) and the Giant Magellan Telescope (GMT). Both are based on segmented optics (although of radically different designs) and will have main mirrors with a total equivalent size of 30 and 25 m, respectively. The TMT will be located near the Keck telescopes in Hawaii and the GMT at Las Campanas in the Atacama desert in Northern Chile. Both projects involve strong international collaboration, including countries such as China, Japan, India and Australia.

However, the largest telescope in the world will be the European Extremely Large Telescope (E-ELT), see Fig. 7.2. Its primary mirror will be composed of 798 hexagonal segments, each

Fig. 7.2 A computer generated image of the forthcoming largest telescope in the world, the E-ELT. The enormous size is indicated by comparison with the trucks in the bottom part of the image (ESO)

1.45 m across, and when fitted together will boast a total diameter of 39 m! The project is a collaboration between fourteen European countries with the probable addition of Brazil. Forerunners to this project were the 100 m Overwhelmingly Large Telescope (OWL) and the 50 m Euro-50 projects, but technological and economical realities scaled down the ambitions of these somewhat. With an estimated price tag of about 1 billion €, the E-ELT will be one of the largest investments ever made in astronomy. The location will be on an Andean mountaintop, Cerro Armazones, at an altitude of 3000 m in the Atacama desert of Northern Chile. This site, as well as the sites for the other two projects, stands as one of the very best locations for astronomy in the world. Its altitude means that the amount of air above the telescope is significantly reduced. In addition, the surrounding Atacama desert is one of the driest places in the world, which facilitates observations further into the infrared part of the spectrum of light. And it is of course a major advantage that the area has clear skies about 350 days every year.

The giant telescopes of the new generation will not only gather much more light than before, but will also be able to obtain sharper images than ever. The resolution in a ground-based telescope is primarily limited by disturbances in the air above the telescope (see Box 7.1). But with the help of new advanced technology in optics, so called adaptive optics (see Box 7.2) it becomes possible to correct for this disturbance. For the E-ELT this should enable images at least ten times sharper (see Fig. 7.3) than the Hubble Space Telescope!

These three new telescopes are colossal scientific projects. With projects involving such extremely advanced technology and enormous investments, it is not surprising that the duration of construction is not exactly known. Currently, the schedules for three telescopes predict "first light" sometime between 2022 and 2025. Whenever they are inaugurated, it is clear that they will revolutionise astronomy in many areas. The exploration of the really young universe, its expansion, investigation of dark energy and the stability of the natural laws are some of the things on the wish list. As with other "first lights" of large telescopes in history, completely unexpected discoveries that we cannot imagine today will probably make the largest headlines.

But what will the extremely large telescopes do for exoplanet research? Well, they will bring detailed observations to an entirely

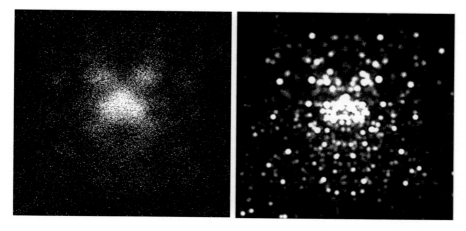

Fig. 7.3 Two computer simulated images of a hypothetical and very distant star cluster (60 million light years). The *left* image corresponds to a 10 m telescope, the *right* to the E-ELT, both using optimal adaptive optics, giving diffraction limited resolution. For this reason already the 10 m surpasses the Hubble telescope resolution, and the impressive potential of the E-ELT is clearly demonstrated (Peter Linde)

new level. Primarily, the focus will lie on exoplanet systems already known from other observations. All the methods previously discussed will be improved enormously. The smallest radial velocity currently measurable for relatively bright stars is about 1 m/s, in itself an astonishing achievement. The E-ELT it is expected to push this to 1 cm/s. At this level of precision it becomes possible to discover even small terrestrial planets. However, at that kind of accuracy many types of motion effects become visible, not least internal motions in the atmospheres of the stars themselves. And of course the minute gravitational influence of a terrestrial planet on its host star will be superimposed on the effects from larger planets in the same system (see Fig. 5.2 in Chap. 5).

In spite of the enormous difficulties involved, direct imaging will be a realistic option, at least for the more nearby exoplanets. This is also the prerequisite for the next important step—to be able to analyse and characterise the properties of terrestrial planets. The combination of incredible light gathering capacity of the E-ELT with capability to obtain extremely sharp images will, for the first time, enable a spectroscopic analysis of the very faint light from an Earth-like exoplanet. This means that searching for the biomarkers previously discussed will become possible (Fig. 7.4).

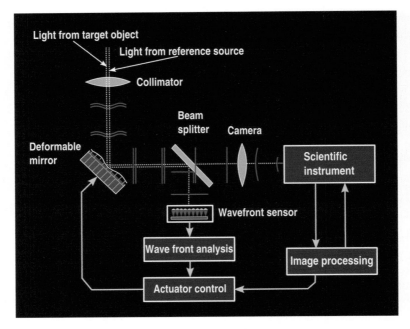

FIG. 7.4 The principle for adaptive optics. A deformable mirror is capable of rapid corrections allowing for cancellation of the disturbances due to atmospheric turbulence. The mirror is controlled by advanced electronics and mechanics, which in turn get information from a wavefront sensor using as reference a longer wavelength light source (Peter Linde)

Box 7.1: A Sharp-Eyed Telescope

The resolution of the image from a ground-based optical telescope is mainly dependent on three factors: the diameter of the mirror, the instrument's optical and mechanical quality, and the detrimental influence of the atmosphere. The most fundamental limit is given by nature itself, as the very wave-nature of the light imposes a limit on the resolution. Mathematically, this is described by the formula $R = 1.22 \cdot \lambda/D$, where R is the resolution expressed in radians (a measure of angle), λ the wavelength of the light in metres and D the size of the mirror in metres. This formula describes what is known as diffraction limited resolution. The smaller value of R, the better. This means that the resolution is inversely proportional to the size of the lens or the mirror of the telescope—the larger it is, the smaller the details can be discerned. It also means that the resolution is worse for longer wavelengths. However, to achieve this theoretical limit, very high demands must be put on the optical figuring of the mirror shape. As a rule of thumb, 1/20th of the relevant wavelength is typically used. For yellow light this corresponds to a precision of 25 nm, i.e. 25 millionths of a millimetre! For a large telescope, similar demands are necessary for its mechanical quality.

However, the third factor is the greatest challenge for a ground-based telescope. Atmospheric turbulence causes a strong deterioration of the image quality. For a large telescope this can amount to a factor of fifty times resolution decrease. Interestingly enough, development of a new high-tech method is now under way (see Box 7.2) to battle this problem. Known as adaptive optics, this technology makes it possible to remove the effects of the atmosphere, allowing the theoretical optical limit to be achieved.

For radio telescopes, atmospheric turbulence is unimportant. Instead, the resolution for radio telescopes is limited by the much longer wavelengths used, about a million times longer than for those of light. On the other hand, radio telescopes can be very large and even connected together with long baselines in between. This gives a very large value for D in the above formula. In this way, radio telescopes can achieve a very high resolution (see Fig 2.6).

Box 7.2: What Is Adaptive Optics?

In an astronomical context, atmospheric turbulence is a small-scale motion, driven by temperature variations. The density of air varies with temperature, and this causes variation in the refractive index of the air. The effect on starlight passing through this medium can be described as an aberration of the incoming plane wave front. These disturbances can be measured using a special detector. Since the turbulence is rapidly changing, this must be done frequently, preferably about 1000 times per second. A reference star needs to be observed to obtain this information. However, during a 1 ms observation, not many photons are available, unless the star is sufficiently bright. As an alternative, an artificial light source can be created using a powerful laser to illuminate a sodium layer near the top of our atmosphere. Through excitation the sodium emits light and appears as a star, affected by turbulence.

Adaptive optics works by using the measured information to equally rapidly affect the shape of a deformable mirror in the optical path of the telescope (see Fig 7.4 for an schematic overview). Such a mirror changes its shape by mechanical interaction with many small actuators, located beneath the mirror. The result is a reconstructed plane wave form which allows the resolution of the telescope to reach the theoretically possible, only limited by diffraction. For a large telescope, this typically means an improvement in resolution of 10–100 times.

To Measure a Coin on the Lunar Surface

Astrometry has always been a cornerstone in the understanding of the universe, from earlier times right up to today's front-line technology. Tycho Brahe was the foremost observer of his time, and with his observations Kepler could decide how the planets moved around the Sun. As we described in Chap. 1, by making positional measurements of the highest possible quality it became possible to measure longer distances and create a clearer understanding of the structure and size of the universe. During the nineteenth century, astrometry was very popular, constituting one of the major disciplines within astronomy. Many observatories of the time installed a special instrument, a so called meridian circle, which was used to carry out high precision measurements. However, the previously discussed atmospheric turbulence limited what was possible to measure from the ground. The discipline fell into some discredit for some time until the event of space based investigations became realistic. Clearly, it would be a great advantage to make such observations from space. In modern times, astrometry plays a very important role. In 1989 the European Space Agency (ESA) launched an epoch-making satellite named Hipparcos. It contained a telescope specially designed for positional measurements, albeit with a mirror size of only 29 cm. Although the satellite did not enter the intended circular orbit Hipparcos nevertheless became a great success.

In order to extract the final results from the data, an enormous effort in terms of computations was necessary. This herculean task fell to about a hundred astronomers, in two independent groups. During 1997, the first results were published in the form of a catalogue with data for the positions, distances and proper motions of about 100,000 stars. The positional accuracy achieved was about half a millisecond of arc. That is the angle that a one metre object would subtend if it was placed at surface of the Moon! Three years later a supplementary catalogue containing 2.5 million stars was published. Together with other measurements of radial velocities of the stars it become possible, for the first time, to obtain a clear three-dimensional understanding of how the stars were distributed in the neighbourhood of the Sun. In turn, this information revolutionised many parts of astronomy, but we would digress too far if we discussed this in detail. Suffice to say that the first rung on the astronomical distance ladder was now very firm, and shed much light onto the finer astrophysical mechanisms of stars. It is also interesting to note that no exoplanets were yet known at the time of the Hipparcos observations.

Based on this success ESA has now gone further by deploying a successor. The satellite, named Gaia (Fig. 7.5), was launched in December 2013, setting an even more ambitious bar. The aim is to measure a billion stars, with a precision of about 10 μm of arc. This is smaller than a quarter coin seen at the distance of the Moon! Simultaneously, Gaia is making measurements of the brightness and colour of the stars. The scientific impact of these observations will be immense. While the main object of the mission is to survey a significant part of the Milky Way, there will also be a host of interesting side effects. At close range to our own solar system, researchers expect to find about 100,000 new minor planets.

FIG. 7.5 The astrometric Gaia satellite was launched in 2013 (ESA/C. Carreau)

Far out in the universe Gaia is expected to study 100,000 super novae in distant galaxies and, through this, shed light on the new theories about the accelerated expansion of the universe and the dark energy. At the very edge of the observable universe, Gaia will study about half a million quasars.

But for our purposes the most exciting part of the Gaia satellite observations is the fact that it should discover a large amount of exoplanets. We mentioned in Chap. 5 that the Sun waggles in its position due to gravitational influence from the planets, mostly from Jupiter. This kind of effect can be measured by Gaia out to a distance of least 1000 light years. However, the limited lifetime of the satellite also limits the length of time that observations of a Jupiter-type planet can continue. The life span is expected to be 5 years, compared to the orbital period of Jupiter, which is 12 years. Within a radius of 150 light years it is estimated that all Jupiter-type planets with an orbital period of 1.5–9 years will be identified. Gaia will also detect exoplanets using the transit method since it carries instruments for high precision photometry. All in all, the scientists expect to find tens of thousands of new exoplanets. Even if the Gaia observations are not sensitive enough to detect terrestrial planets, they will doubtless lead to a huge knowledge base into which researchers can dig for other investigations. For example, the foundation for the next generation of exoplanet satellites will be laid by the extremely accurate observations of stars within the closest 100 light years. Among these, suitable candidates for life-bearing planets can be selected for detailed scrutiny. Gaia is currently operating nominally, although scattered light levels are higher than expected. In September 2014, Gaia discovered its first supernova in a galaxy, half billion light years away.

The James Webb Space Telescope

In 1993 the Hubble Space Telescope (HST) at last began to deliver sharp images. It was about ten years behind the original schedule. Since then it has become a fantastic success both scientifically and in terms of publicity. A visiting space shuttle crew made the last upgrade and repair in 2009. But since the retirement of the space shuttles there are no more possibilities for repairs. Nevertheless, the current status (2015) of the telescope along with its instruments is good, and plans are being discussed for continued service until about 2020. The nearest planned successor to the Hubble Space telescope will be the James Webb Space Telescope (JWST), now expected to launch in late 2018. While it will not have the HST capabilities in the shorter optical and ultraviolet parts of the spectrum, it will be far superior in the infrared. Studies of the infrared part of the spectrum are considered fundamental for many important astronomical issues. Much of the light that originates from the young universe is best observed in the infrared, due to the cosmological red shift. The creation of new planets takes place primarily at low temperatures, which means infrared radiation. In addition, intervening interstellar dust is more transparent to such radiation, allowing penetrating observations of otherwise obscured objects. These are the main reasons behind the construction of the JWST.

The James Webb Space Telescope (Fig. 7.6), named after a previous NASA director, is an ambitious project. The route to space

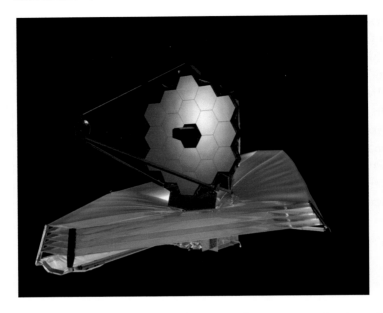

FIG. 7.6 The James Webb Space Telescope. The segmented mirror is protected with a heat shield consisting of several layers (NASA)

for the telescope has met with many obstacles and delays, in a way similar to many other space projects. The original cost for the project was estimated at $1.5 billion with a planned launch for 2011. A seven year delay and a new cost estimate at $10 billion put the project in serious jeopardy, and in 2011 the US senate decided to cancel it. A wave of protests among astronomers and some politicians reopened the discussion and the decision was revoked in November 2011, allowing the project to continue.

JWST will be launched into a rather peculiar orbit. As is the case with the Gaia satellite it will orbit around a Lagrange point, in the neighbourhood of the Earth, but at a distance four times that to the Moon. The telescope will be innovative in many ways. It will be the first in space with a segmented mirror. Eighteen segments will be unfurled after the launch to form a 6.5 m mirror. Since the JWST will be working in the infrared, it is imperative that the craft is kept as cold as possible. A multi-layered heat shield will protect it against solar radiation. The craft is expected to maintain a temperature just forty degrees above absolute zero (i.e. about –230 °C). And like the Hubble telescope, Europe is participating in the project with fifteen percent share. This includes the construction of one of the main instruments and, not least, the actual launch will take place in French Guyana with the ESA Ariane 5 rocket. This launcher has a five metre diameter, and to encase JWST in there has been compared to building a ship in a bottle...

JWST incorporates unique capabilities to assist in the exoplanet research. As we already noted, the telescope is perfect for studies of the early phases of planetary system formation, for example the accretion disks which often exist around young stars from which planets are created. Temperatures in these places are rather low and the matter emits heat radiation of a kind that JWST can detect. But direct studies of exoplanets will also be possible. Above all, JWST will function as an excellent successor to the Spitzer space telescope. Spectroscopic studies of the atmospheres of the exoplanets will become much more detailed, and we can look forward to new exciting discoveries.

Future Space Initiatives

Although NASA has long has played a dominant role in space-based research, several European projects have become successes. We earlier mentioned the Hipparcos satellite, which through its uniquely accurate positional measurements helped to set a firmer footing for many fields in astronomy. Now, its successor Gaia is in an excellent position to continue its work. ESA is also undertaking several positive collaborations with NASA. But what follows next? The document "Cosmic Vision" from 2007 presents the European plans for space science during 2015–2025. The four main themes of focus are:

- Under which circumstances are planets created and life initiated?
- How does the solar system work?
- What are the fundamental laws of physics in the universe?
- How was the universe created, and what is it made from?

ESA uses a system of space projects with three different levels, small, medium and large projects.

CHEOPS was selected in 2012 as the first small class space mission. Slated for launch in 2017, the mission aims to bring an small 30 cm telescope into an orbit of about 800 km altitude. For the planned mission duration of three and a half years, CHEOPS will examine transiting exoplanets at known bright and nearby host stars. Its main goal will be to accurately measure the radii of the exoplanets for which ground-based spectroscopic surveys have already provided mass estimates. Thus, since it will primarily study already identified systems, it will optimise efficiency by observing only during transit phases.

In the medium sized group three projects have been selected. One is the Solar Orbiter, for detailed studies of the Sun. Another is the Euclid, mainly aimed at studies of dark matter and dark energy. Euclid has an expected launch date of 2019. Among other things, it will study effects of gravitational lensing. This means that it will also look out for exoplanets, including terrestrial ones and even smaller. The third ESA choice is the most interesting one from the exoplanet standpoint.

PLATO (PLAnetary Transits and Oscillations of stars) seeks to identify and study thousands of exoplanetary systems, with an emphasis on discovering and characterising Earth-sized planets and super-Earths in the habitable zone of their parent star. It will use 32 special cameras, covering a field about twenty times larger than that of Kepler's. PLATO will differ from the COROT and Kepler space telescopes in that it will study relatively bright stars (between magnitudes 4 and 8) making it easier to confirm exoplanets using follow-up radial velocity measurements. Plan calls for a launch in 2024, with an operational time of six years.

Recently JUICE (Jupiter Icy moons Explorer) was selected as the first large-scale enterprise. JUICE entails a journey to Jupiter beginning in 2022. Eight years after launch, and after a series of complicated gravity assist manoeuvres involving Venus and the Earth, the spacecraft will enter the Jovian system in 2030. Three of the Galilean moons will be investigated in detail, Callisto, Europa and Ganymede. These may have liquid oceans beneath their ice crusts and could potentially harbour life. A possible collaboration with the Russian space agency, Roscosmos, could also result in including a Ganymede lander. As we now know, Jupiter-type satellites are common among the stars, and it is fundamental to understand them and, in particular, their moons. A Jupiter planet in the habitable zone around a star is itself probably not a suitable place for life, but if it has large moons these may be ideal.

Meanwhile, NASA is making plans for at least three space-craft devoted to exoplanet research. In 2016, spacecraft Juno will arrive in the Jupiter system, primarily to measure the properties of the Jovian atmosphere. A much more ambitious project is the Europa mission, which will conduct detailed reconnaissance of Jupiter's moon Europa and investigate whether the icy moon could harbour conditions suitable for life. The mission plan calls for the spacecraft to be launched to Jupiter in the 2020s, arriving in the distant planet's orbit after a journey of several years. The spacecraft would orbit the giant planet about every two weeks, providing many opportunities for close flybys of Europa.

A successor to the Kepler satellite is also planned. TESS (Transiting Exoplanet Survey Satellite) is designed to carry out the first space-borne all-sky transiting exoplanet survey. It will be equipped with four wide-angle telescopes and associated CCD detectors. The primary mission objective for TESS is to survey the brightest stars near the Earth for transiting exoplanets. Current plans calls for a launch in 2017.

The Wide-Field Infrared Survey Telescope (WFIRST) is a proposed NASA space observatory designed to perform wide-field imaging and spectroscopic surveys of the near infrared sky. The current Astrophysics Focused Telescope Assets (AFTA) design of the mission makes use of an existing 2.4 m telescope. WFIRST-AFTA will study essential questions in both exoplanet and dark energy research. A coronagraph instrument will allow direct imaging of exoplanets and debris disks, and a photometric survey will use microlensing to look for exoplanets. The aim is for a launch in the mid 2020s.

New Instruments on the Ground

Rapid development is also taking place on the ground. As we have earlier seen, it is not just about building powerful telescopes for surveying projects and for special studies. You also need to build advanced instruments to analyse the light. Based on recent experiences successors to HARPS and similar spectrographs shortly will be mounted on the very largest telescopes.

In fact, in 2015, SPHERE (Spectro-Polarimetric High-contrast Exoplanet REsearch) began operation at one of ESO's 8 m VLTs. The primary science goal of SPHERE is imaging, low-resolution spectroscopic, and polarimetric characterization of extra-solar planetary systems. The instrument design is optimised to provide the highest image quality and contrast performance in a narrow field of view around bright targets that are observed in the visible or near infrared.

Simultaneously large surveying projects, both with small and medium-sized telescopes are under development. There is still room for improvements to intensify studies using the microlensing and transit methods. The ambition is to monitor even larger numbers of stars and to do it more often. One such project is the Next-Generation Transit Survey (NGTS) which consists of twelve individual but coordinated telescopes. NGTS can be considered as a further development of the previously discussed WASP project. In early 2015 it saw "first light" beginning its operation located at Paranal in the Andes of Chile, just beside ESO's Very Large Telescope. The set of small 20 cm telescopes altogether cover almost 100 square degrees on the sky. The observations will be focused on relatively bright stars with the goal of finding planets smaller than Neptune, typically super-Earths with sizes 2–5 times that of the Earth.

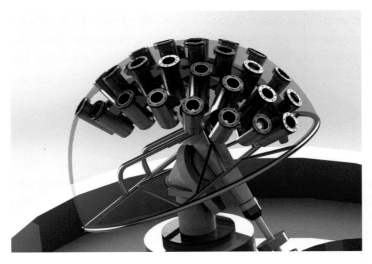

Fɪɢ. 7.7 The innovatively designed Evryscope, containing 27 small telescopes, carried on the same mount (Nicholas Law)

There are several more ground-based projects soon beginning observations. CARMENES is Spanish-German collaboration to use a new high-resolution spectrometer on a 3.5 m telescope to detect low-mass planets in the habitable zones of red dwarfs. The MEarth Project consists of two robotically controlled observatories, both containing a set of eight medium-sized telescopes. The Northern observatory is located on Mount Hopkins, Tucson, Arizona, the Southern at the Cerro Tololo observatory, near La Serena, Chile. A similar project is the Italian APACHE, operating from the Alps.

An interesting new technology development is the Evryscope Project, located at Cerro Tololo (see Fig. 7.7). It consists of an array of 27 individual small telescopes mounted on a common mount. It uses CCD detectors with a total of 780 million pixels. Among many possible applications, this instrument could detect transiting giant planets or, around nearby red dwarfs, rocky planets in the habitable zone.

All professional observatories nowadays are located in rather exotic places, but the fact is that some even more exotic places remain to be explored. One such site is the Antarctic, where the infrastructure currently is still weak but growing. An observatory there would only be capable of monitoring the sky about half a year at a time, but on the other hand this would mean half a year of dark conditions the other half, making it feasible to perform continuous observations during long periods of time. In spite of otherwise harsh conditions, the atmosphere is unusually stable and dry there. The so called katabatic temperature-driven winds need be avoided but tests have shown that this is possible at 10–15 m above the ground.

Certainly the creativity and imagination of scientists will continue in directions that we cannot easily predict. And there are known techniques that will certainly be utilised and perfected in the near future. One example is measuring the polarisation from of the light from the exoplanets. This is expected to reveal information about atmospheric conditions and also has a potential for studying surface structures, such as vegetation. Another example is taking advantage of adaptive optics in microlensing observations. Improved resolution is key to improved detections.

A third example would be to try to detect the radio radiation that ought to be emitted from exoplanets experiencing powerful thunderstorms or aurora-type phenomena. New radio astronomy observatories like the gigantic Low Frequency Array (LOFAR, in Europe), the Atacama Large Millimeter/submillimeter Array (ALMA, in South America), and eventually the Square Kilometre Array (SKA, South Africa/Australia) stand ready to meet that challenge. A discussion of these instruments is given in Chap. 11.

Vision for the Future: An Interstellar Radar

We have seen how astronomical observations are all about finding ways to carefully, patiently and shrewdly collect light that has been on its way towards Earth for many years. Within the solar system we have the opportunity to do in situ observations by sending space probes or even manned space flights. To physically visit exoplanets is, however, currently clearly impossible, even though we will look more closely at such ideas in Chap. 12. But there may be yet another possibility, namely to actively study exoplanets with radar. Radar studies of objects in the solar system have already been carried out for decades. By transmitting radar pulses in different ways to study their reflection, it is possible to obtain detailed information. Radar echoes have been successfully received from as far as the rings of Saturn and its moon Titan.

The American scientist and electronics expert Louis K. Scheffer recently presented the concept of building an interstellar radar system. He believed that this technology is within reach now, both technically and economically. The problem is that interesting distances (10–15 light years) are about 100,000 times longer than distances to Titan, so such a system needs to have gigantic dimensions. Scheffer estimates that an area the size of Hawaii (10,000 km^2) would need to be filled with receiving dishes, and an area about ten times larger (about the size of Iceland) would be needed for the transmitting portion. Suitable isolated and radio quiet areas can be found in the inner parts of Australia or in Africa. Scheffer estimates that the energy needed for the transmission, about 10 Terawatts—corresponding to about the output of 10,000 nuclear power plants—is already available since locally infalling solar energy could be used. He estimates the cost of such an undertaking constructed during a time period of twenty years would be comparable to military spending during the same period.

With interstellar radar it becomes possible to study in detail hundreds of exoplanets within the nearest twenty light years. Naturally, one would need to wait for twice the corresponding time to get any results but this is significantly faster than an interstellar expedition would require. In any case, such a radar system would pave the road for future in situ studies.

The Legacy of the Kepler Satellite

While researchers wait for new radical measurement methods and results, the Kepler satellite still is the most important source for new information, and is expected to remain so a few more years. Nearly 4700 candidate planets need further study and, as we have seen, results are coming in at a steady rate. In addition, there are estimates of what may be hiding in the vast material.

It is expected that about 80 % of the candidates will turn out to be actual exoplanets. Perhaps a hundred of them will be about Earth's size. About fifty will lie in the habitable zone, and it is conceivable that some of them may look very much like Earth. But detailed studies of these will have to wait a bit into the future since they all lie at large distances from us.

8. Signs of Life from Earth

The idea of sending messages to other worlds is not new. As early as the 1820s the German mathematician Carl Friedrich Gauss suggested that a giant triangle should be cut in the Siberian forests, to be replaced with wheat. This, he said, ought to be possible to observe from Venus or Mars, the planets then believed to have life. Twenty years later the astronomer Joseph von Littrow proposed an alternative along the same lines. An enormous circular canal was to be dug out in the Sahara desert. It would have a diameter of 30 km and be filled with kerosene, then a technical novelty. By starting a fire at night, the burning circle would display the existence of intelligent life on Earth to hypothetical Venusians or Martians. However, none of these grand schemes were ever realised. It would take more than a century before such ideas again were taken seriously. And at that time nobody considered that either Venus or Mars possessed inhabitants.

The Space Race Begins

On October 4, 1957 the space age began. The world was taken by surprise when the Soviet Union announced that they had launched a satellite, Sputnik 1 (see Fig. 8.1), in an orbit around the Earth. Soon the announcement could be verified from independent sources. Radio amateurs all over the world easily caught the beeps from the transmitter of the satellite. This was the start of a space race between the USA and the Soviet Union that essentially lasted until July 21, 1969, when Neil Armstrong took his historic first steps on the Moon. During this period, every new manned space expedition generated great excitement in the media. In retrospect it appears highly remarkable how rapidly and successfully the space flight technology developed during these twelve short years.

© Springer International Publishing Switzerland 2016
P. Linde, *The Hunt for Alien Life*, Astronomers' Universe,
DOI 10.1007/978-3-319-24118-0_8

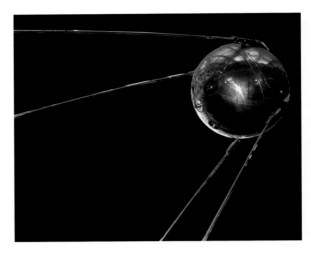

Fig. 8.1 The Soviet satellite Sputnik 1 which initiated the space race (NASA)

The explanation lay in the political prestige and the military importance represented by the conquest of space. The usefulness of Earth-orbiting satellites of various kinds, both for military and civilian purposes, became obvious. Surveillance and communication reached entirely new levels. But by the end of the 1970s resources for further space research were drastically cut. The political incentive had faded after the end of the space race. Instead, at a considerably slower rate, cooperation took its place. A couple of decades later, this new spirit of collaboration resulted in the construction of the International Space Station (ISS). Since November 2, 2000 there have always been humans in space, a not well known, but nevertheless historical, fact.

The space race of the Cold War, which mainly focused on manned space flights, triggered a spin-off: the 1970s saw a number of more or less successful unmanned expeditions through our own planetary system. Space probes from both super powers were at an early stage sent off towards Mars and Venus (cp Chap. 4). But the exploration of the outer planets became an entirely American affair. In 1972 Pioneer 10 set out for Jupiter and its system of moons. Pioneer 11 followed the year after. After its encounter with Jupiter, it went on to study Saturn. Both spacecraft were accelerated to higher speeds using the gravitation and movement of Jupiter as a stepping stone, a procedure we know as gravity assist. In this way they reached a sufficient velocity to overcome the gravitation of the Sun, enabling them to move out of our solar system. Thus, they became our first ambassadors to the stars, and through them humanity has left at least some traces for the very distant future.

The success of the Pioneer probes was followed by an even more ambitious project, known as Voyager. In 1977 two new spacecraft were launched, Voyager 1 and 2. The primary targets again were the gas giants Jupiter and Saturn. The spacecraft both succeeded excellently. Very detailed images of the planets and their moons led to several new discoveries. The technique to use the gravity assist for changing speed and direction was now well developed. Voyager 2 took full advantage of this. After its Saturn visit, it was redirected towards Uranus and subsequently towards Neptune. Voyager 2 successfully reached Neptune after a twelve year voyage. The two Voyager craft also left the solar system and are now on their way toward eternity.

The world would have to wait until 2006 before NASA decided to launch another spacecraft, New Horizons, towards Pluto, considered to be the last and outermost planet in the solar system. Along the way, having studied in detail Jupiter (see Fig. 8.2) and passing the Pluto system in July 2015, it will become the fifth spacecraft to leave the solar system. Since 2006 Pluto has, somewhat controversially, been degraded from planet to dwarf planet by the International Astronomical Union. Dwarf planets belong to a new category of objects in the solar system, and many more have been discovered (see Fig. 2.1, Chap. 2). New Horizons is not carrying an official and dedicated message from humanity. It does contain a small cannister with part of the cremated remains of Pluto discoverer Clyde Tombaugh. A private initiative is also trying to convince NASA to use the memory banks of the craft to upload and store a greeting, after the main objectives of the mission have been completed.

Some facts about the interstellar space probes are given in Table 8.1 and their current positions are shown in Fig. 8.3. We notice that their speed relative to the Sun is about 10–20 km/s. In daily life this stands out as exceedingly fast, but for an interstellar vehicle it is hardly even a snail's pace. It corresponds approximately to 1/20,000 of the speed of light. This is equivalent to saying that they need 20,000 years to complete a single light year of travel.

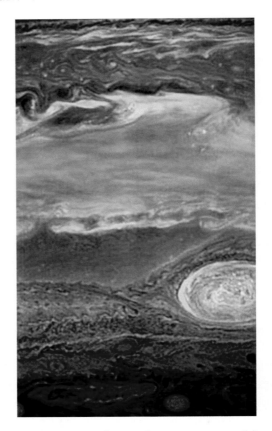

Fig. 8.2 Details in the atmosphere of Jupiter, imaged by the spacecraft New Horizons (NASA/ESA/A. Simon Miller)

Table 8.1 Basic facts (2015) for the five interstellar spacecraft

	Pioneer 10	Pioneer 11	Voyager 2	Voyager 1	New Horizons
Date of launch	Mar 3, 1972	Apr 6, 1973	Aug 20, 1977	Sep 5, 1977	Jan 19, 2006
Distance from Sun (AU)	113.8	93.1	108.5	132.0	32.9
Light time (hours)	15.9	12.8	14.9	15.9	4.2
Speed relative to the Sun	12.0	11.3	15.4	17.0	14.5
Constellation position	Taurus	Scutum	Phoenix	Ophiuchus	Sagittarius
Still working?	No	No	Yes	Yes	Yes

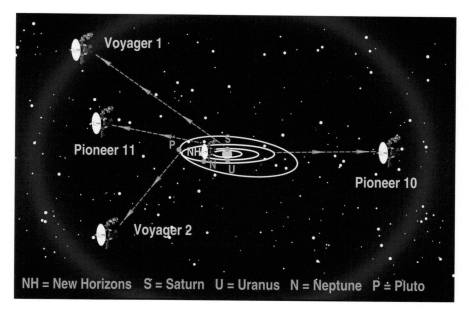

FIG. 8.3 The positions of the interstellar space probes in 2015. The plane-tary orbit of the outer solar system and the so called heliopause are marked (Peter Linde)

Pioneer's Plaque

The first message intended for future eyes was constructed for the two Pioneer craft. Launched in the beginning of the 1970s they would, for the first time, explore the outer parts of our solar system and then leave it. No one could predict with any certainty just how long a time they would remain in their slow interstellar voyage; reasonable estimates pointed to millions of years. The space between the stars is not completely empty, but on the other hand there did not seem to exist any major obstacles that would affect the spacecraft. A certain amount of erosion from interstellar dust could be expected, but at an incredibly low level. Odds favored that the Pioneers would prove to become a future testimony to the existence of mankind.

A historical message was not planned from the very beginning, but the idea came just before the first space probe, Pioneer 10, was to be launched. In three short weeks a proposal was put forward under the guidance of astronomers Carl Sagan and Frank Drake.

It took the form of an engraved plaque, 15×23 cm, 1.3 mm thick and made in aluminium with a thin coating of gold. The engraving itself was 0.4 mm deep.

What should the engraving show? The authors themselves admitted that the short time available did not allow for a deep analysis of what would be a suitable greeting. Every type of message would inevitably be coloured by human senses and human ways of thinking. Nevertheless, an information-dense greeting was crafted eated with an emphasis on humanity and her origin.

How Is the Plaque Interpreted?

The interpretation of the plaque assumes that a receiving civilisation has advanced to at least the same technological level as ourselves. It also assumes that such a civilisation shares similar physical and mathematical basic relations and concepts. For example, numbers can be expressed in binary form. This number system, which is also used in our computers, is the simplest possible numerical expression, and has the advantage of containing only two symbols. These can take any form as long as they are different. In a computer they are simply represented by the conditions on and off. Another example of binary code is the Morse alphabet, which consists of short and long signals.

Figure 8.4 shows the plaque. At upper left there are two large circles with some small lines. These are intended to define a universal length unit. The symbol represents two well-known energy states of the simplest and most common basic element in the universe, namely hydrogen. If the only electron of the hydrogen atom changes its spin condition to a lower state it will emit radio radiation with the wavelength 21 cm. Its frequency is 1420 MHz, which can be interpreted as a time unit (about 0.7 ns). These facts should be well-known all over the universe.

The right part of the plaque contains figures of a man and a woman, rendered with the spacecraft itself in the background. To the far right there are two horisontal, short, lines. In between these, a binary number corresponds to 8. 8×21 cm is 168 cm, the height of the women. This can also be compared to the drawing of the spacecraft, whose real dimensions obviously are known to the decipherer.

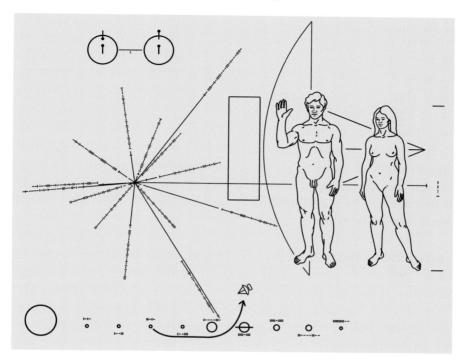

FIG. 8.4 The plaque carried by the Pioneer spacecrafts, with information to the future (NASA)

The rest of the message is about the spacecraft's, and therefore humanity's, origin. Here we encounter a problem that we will discuss more at a later point: Is it reasonable and harmless to reveal our location in the Galaxy to an alien civilisation? Most people would argue that the time perspective until such a hypothetical discovery of the spacecraft is so large that, at that point, it hardly makes any difference.

At the bottom there is a schematic illustration of our solar system. It is not to scale, but the relative distances between the planets are given in binary numbers. Pioneer's initial path is marked and it has its antenna pointing towards the third planet, the Earth. Today, we know that planetary systems are commonplace among stars, and the intention of the illustration is simply to distinguish it from other, possibly nearby, systems.

The position in the Milky Way of our solar system is more accurately given by a kind of map on the middle left part of the plaque. A number of lines, intersected by binary numbers, point in various directions. In fact, they give the directions to a set of 14 different pulsars. Pulsars are celestial bodies that can be formed during supernova explosions, and they are rapidly rotating radio sources. The first one was discovered in 1967 (see Chap. 11). The radio transmission is detected as short, intensive pulses, much like a lighthouse. The time between two pulses, the period, can be determined to great accuracy. They are different for different pulsars but are usually in the interval 0.1–3 s. It is also known that the rotation of pulsars slows down somewhat as the millennia passes, due to energy losses. The period increase can be as small as one second over a 10 million year time period. This also makes them usable as galactic clocks.

The binary digits on the lines of the plaque are large numbers, corresponding to ten digit numbers in our ordinary decimal system. This suggests a characteristic that can be determined to very high precision. The only possible property is the period of a pulsar. The lines provide the approximate position and distance to each respective pulsar. Put together, this information allows an observer to determine the origin in time and space of the spacecraft. A complication would of course be that both the Sun and the pulsars themselves may have moved with respect to their original positions.

The two Pioneer probes became scientific successes in their exploration of the planets Jupiter and Saturn. Thanks to them, both planets were imaged at close range for the first time. Afterwards, the Pioneers continued out of the solar system, transmitting valuable data about interplanetary space. In 1997 the Pioneer 10 project was officially terminated. In 2003 the final, very weak, signals were received from the probe. It then continued its voyage towards the stars in the constellation of Taurus. In about two million years it is expected to once again come close to a star.

The contact with Pioneer 11 was terminated in 1995. It is on its way to the constellation of Scutum where, in about four million years time, it is expected it to come close to a star again.

Voyager's Golden Record

In 1977 two more advanced Voyagers spacecraft followed in the footsteps of the Pioneer spacecraft. They were also destined to eternity, after having studied the outer parts of our solar system.

But this time, the preparations for putting together a message for a future alien civilisation were somewhat better. Again, Carl Sagan supervised the project. In the book *Murmurs of Earth*, Sagan and co-authors describe in detail the fascinating story of how it was done. The level of ambition was now quite a bit higher. The team wanted to make a serious attempt to characterise humanity, its origin, culture and technological level. Sagan put together a working group, including well-known astronomers and a professor of philosophy and social thinking. They consulted with big names in science fiction literature, like Arthur Clarke, Isaac Asimov and Robert Heinlein. It was decided that a suitable medium would be a kind of gramophone record, manufactured in a durable metal (see Fig. 8.5). On this record, both sound and images could be imprinted. The mechanical grooves that contained the information would survive for much longer than any type of magnetic storage available in the 1970s. Our modern CD record, based on laser burning and use of digital techniques, was not introduced until the beginning of the 1980s.

Now the question was how to make suitable choices for the content. After much discussion, the group decided that the record would only display the good aspects of humanity; no violence or war would be illustrated. At the same time, any obvious reference to religion was removed, with the motivation that it would be impossible to include all the monuments and symbols of all the religions.

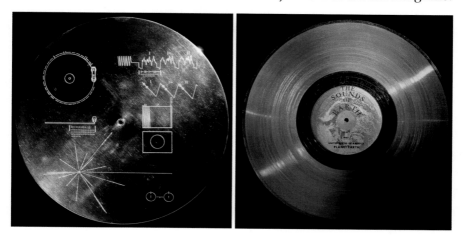

FIG. 8.5 The golden record carried by the Voyager spacecrafts. To the *left* is seen the cover with basic instructions and information, to the *right* the actual record (NASA)

Sagan now used his large network to get proposals for the contents. The task was delicate; it was almost unavoidable to be criticised after each choice or each exclusion. The Pioneer plaque, with its limited capacity for information, had briefly described the origin of humanity from a purely scientific point of view. Now the aim was to convey not only what we know and how we think, but also how we feel. To include music, therefore, was a natural step. Someone suggested including the complete works of Bach, but, as the proposer himself pointed out, "it would be to boast…"

Voyager's Sound Recordings

After a careful weighing of many suggestions, the group defined a list of 27 musical pieces, containing almost 90 min of recording. The result constituted a mix of world music, with a slight over-representation of Western classical music. Among the represented composers were names like Mozart, Stravinskij, Beethoven and—Bach. Chuck Berry's "Johnny B. Goode" was chosen to represent more modern music. Also incorporated were music recordings from many different cultures. Examples include aborigine music from Australia, pygmy music and music from Peruvian Indians. In Appendix B at the end of the book there is a complete list of the musical content.

Another type of sound recording came from animals and nature. Thunderstorms, streaming water and rain are examples, as are the characteristic animal noises of a dark African jungle. In a more Northern setting, a wolf is howling in a storm and dogs are barking. The recording also contains sounds from human activities, everything from a beating heart to the wailing of an operating sawing machine. Machine-generated sounds from various industrial processes and epochs are also represented, steam engines, cars, airplanes and much more.

Some recordings are greetings. Former United Nations Secretary-General, Kurt Waldheim, sent a greeting representing all peoples on Earth. Additionally, a number of delegates from the UN committee of outer space got a chance to send a short greeting in their own language. However, when the recordings were to take place it turned out that the ambassadors' concept of "a short greeting" differed considerably from that of the organisers. The solution was to make an edited mix where fragments of the speeches were interspersed. A notable exception was the recording from the Swedish representative. He chose instead to recite a famous and moving poem, "Visit To An Observatory" by the Swedish Nobel laureate Harry Martinson. Sagan was so impressed by this that it was kept in its entirety.

To get a more representative coverage of human languages it was also decided that a short phrase of greeting would be recorded in as many languages as possible. However, time was now short and Carl Sagan turned to the linguistic departments of his university, the Cornell University. In this way, greetings were finally recorded in 55 different languages.

Voyager's Images

The record holds 122 pictures. By modern standards, their technical quality is modest; the resolution resembles that of a normal TV image. Great care was taken in choosing the images. The first ones are purely technical and intended to form a basis for interpreting the following images. Our numbering system and our measurement units are introduced. These are subsequently used to give an idea of scales and timing aspects. An overview of the solar system defines our location.

Then, an image explains the fabric of life on Earth, describing DNA molecules. Human birth and anatomy is shown in a sequence of images. When Carl Sagan asked NASA for approval of the images they said OK to all of them except one. The rejected one illustrated a naked man and women holding hands. The woman was a few months pregnant. NASA had received criticism in the past for the stylised images on the Pioneer plaque. Now, NASA refused completely. The solution was a compromise where the images were changed into silhouettes with the embryo specially marked in the womb of the women. Subsequent images show animals and nature and various expressions of cultural and human activities. The last ones emphasise technological achievements of various kinds.

Putting together material for the golden record was not always easy, especially against the backdrop of the cold war. Sagan wanted

Fig. 8.6 Voyager 1 on its way into eternity. The disk is mounted on the outside of the craft. In the background is shown the constellation of Ophiuchus, the red square shows the direction of travel in the sky (NASA/JPL-Caltech/Peter Linde)

to include a famous Russian folk song but thought it wise to first consult a Soviet colleague to find out if it was acceptable. Long after the deadline had passed for an answer came a reply, which probably originated from very high political levels. "A Night in Moscow" was the word. But at that time Sagan had already resigned and instead the choice was "Chakrulo", a Georgian folk song.

Contact with the Voyagers will be maintained for another few years. Voyager 1 is moving in the direction of the Ophiuchus (see Fig. 8.6) and may pass into the neighbourhood of another star in about 40,000 years. A similar undisturbed future awaits Voyager 2, on its way to the constellation Phoenix.

Have We Revealed Our Existence?

When the spacecraft were equipped with information addressed to a hypothetical future alien civilisation, some worried that Earth could be put in peril if the receiver would turn out to both hostile and superior. However, one should consider the fact that mankind has long been signalling its existence to the universe. First in mind are the radio transmissions that started with the Marconi experiments more than century ago. Since then, transmitters have become much more powerful, capable of reaching all over the world. About 70 years ago the first experiments were made with television broadcast using higher radio frequencies. Parts of these transmissions leaked into space and are now travelling further and further away at the speed of light. So it can be correctly stated that Earth is now in the centre of a steadily expanding globe of radio radiation, with a current radius of about 100 light years. However, a closer analysis shows that the great majority of these transmissions are so diluted that they hardly can be perceived outside our solar system, even by an advanced civilisation. Nature's own laws of physics limit how far such weak signals can be perceived. There are, though, some much stronger signals in the radio noise transmitted from Earth, primarily from powerful radar systems. Such broadcasts are directional and therefore penetrate to much longer distances, even on interstellar scales.

Beyond this, high frequency and energetic radiation, mainly in the form of gamma and X-ray radiation, is on its way into space, emanating from the nuclear bomb explosions made in the 1940s and 1950s. Unfortunately, these carry a signature that should be quite easy for an alien observer to interpret.

As we have already seen in the chapter about exoplanets, in a reasonably near future we foresee ourselves being able to determine whether life exists on some planet in our own neighbourhood. Then it should be a reasonable conclusion that an advanced alien civilisation would be as capable of at least determining the existence of terrestrial life. So it is not inconceivable that we may already have caused some attention. It is, however, sobering to have a look at Fig. 8.7. It illustrates the maximum extent to which any human activity may be observable so far in the Milky Way. The size of the selected square corresponds to about 6000 light years. In the centre of this the Sun is located, the yellow dot represents a diameter of about 200 light years. This is the current maximum reach of human radio signals.

Thus far, five human interstellar spacecraft carry messages that, at least theoretically, can be deciphered sometime in a distant future. The probability of this really happening is of course

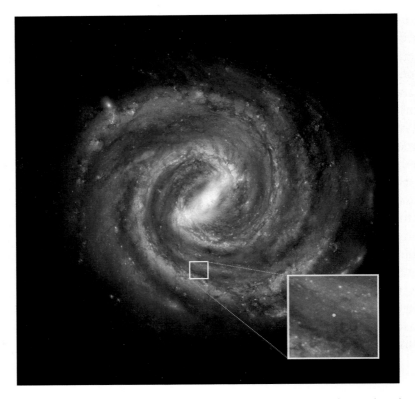

Fig. 8.7 An artist's conception of the Milky Way, as seen from the above. The enlarged area contains our solar system. The yellow dot corresponds to a diameter of 200 light years, the maximum reach of any terrestrial radio signal (Nick Risinger/Peter Linde)

very slim. If you want to be serious about communicating with the universe, the only reasonably realistic method of doing so is a series of powerful and well-directed signals travelling at the speed of light. A number of such messages are, in fact, already on their way right now.

The Arecibo Message

The gigantic radio telescope in Arecibo, Puerto Rico (see Fig. 8.8) was built already in 1960. An entire valley was used as foundation and when it was ready it became (and still is) with its 305 m diameter the largest antenna in the world. However, radio technology

FIG. 8.8 The large radio telescope at Arecibo in Puerto Rico (NAIC/Arecibo Observatory/NSF)

was advancing rapidly and in 1974 it was time for an extensive refurbishment. Since the telescope can also be used as a radar, a 500 kW transmitter had been installed. To celebrate the upgrade it was decided to send a message to the stars. The transmitted message lasted for only three minutes and was pointed in the direction of the giant globular stars cluster M 13 (see Fig. 8.9), which lies in the constellation of Hercules. It is considered to contain at least 300,000 stars and the idea was that somewhere in the cluster there could exist a civilisation with capability to detect and decode the signal. The distance to the cluster is about 25,000 light years, which consequently means that it will take 25,000 years before the message enters the cluster.

What form should such a message take, so that it contains important information and at the same time is suitable for a radio transmission? Again, it was Frank Drake, who—with the help of Carl Sagan and a few students—sat down and did some thinking. The result was to send information formatted as a simple image.

Fig. 8.9 The inner parts of the M 13 the star cluster. The Arecibo message is on its way towards this cluster (ESA/Hubble/NASA)

The image would contain a sequence of radio pulses, and was to be transmitted at a frequency of 2380 MHz, corresponding to a wavelength of 12.6 cm. By slightly alternating the signals in frequency, the two binary numbers 0 and 1 were represented. Exactly 1679 bits were broadcast. This number was chosen because it happens to be the product of the two prime numbers 23 and 73. Correctly put together, the 1679 bits formed an image consisting of 23×73 pixels, containing zeroes and ones. In Fig. 8.10 we see to the left what such a formation may look like. It obvious that the digits are not placed completely at random. In the centre panel the same information is displayed more graphically, with all the zeroes as black and the others strategically coloured.

The resulting short message is rather complex and demands considerable analysis from future inspectors (as well as contemporary ones!) in order to be decoded. The image should be interpreted as

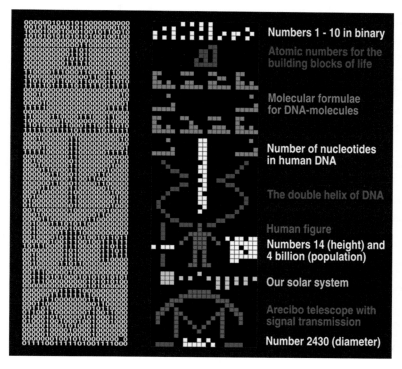

Numbers 1 - 10 in binary

Atomic numbers for the building blocks of life

Molecular formulae for DNA-molecules

Number of nucleotides in human DNA

The double helix of DNA

Human figure

Numbers 14 (height) and 4 billion (population)

Our solar system

Arecibo telescope with signal transmission

Number 2430 (diameter)

Fig. 8.10 The message from the Arecibo telescope. To the *left* its original binary form, in the middle the ones are colour coded for clarity, to the *right* explanations (Peter Linde)

a mixture of binary and graphical information. The emphasis lies on explaining mankind's biological descent and place in the universe, in many ways resembling the philosophy behind the Pioneer plaque. At the top, a small diagram explains our decimal numbering system, where the numbers 1–10 are represented in binary form. These numbers are then used to point at the numbers 1, 6, 7, 8 and 15, which are the atomic numbers for the most important basic elements in living organisms (hydrogen, carbon, nitrogen, oxygen and phosphor). Using this as a basis, the formulas for the molecules in our DNA (see Chap. 3) are given, supplemented by a figure of its spiral structure and a large number representing the number of nucleotides (molecular building blocks) in our DNA. A small human-like figure is flanked by two numbers. The number 14 should be interpreted as the length of the human, using the wavelength of the signal, 12.6 cm, as the measuring unit, which gives 176 cm. The number four billion refers to the number of humans existing when the message was sent. The human figure stands on a small model of the solar system, closest to the third planet, our Earth.

The Sun is to the left and the giant planets to the right (including dwarf planet Pluto). At the very bottom, a simple sketch of the radio telescope itself transmits the message: the number 2430 is interpreted as its diameter, about 300 m, again in units of 12.6 cm.

As the message lasted for three minutes, it thus became three light minutes long. It has now reached about 40 light years out into space and is the most powerful radio beam that has ever been sent from Earth. It is very well focused and for a hypothetical observer in M 13 Earth will by far appear as the strongest radio source in the Milky Way during three short minutes in about year 27,000.

The experiment has, however, been criticized for various reasons. Some claim that it is unwise to so clearly advertise our existence. But again the time scales involved are somewhat mitigating: no reaction of any type can be expected for at least 50,000 years. A remote but possible exception could be civilisations belonging to the few local stars lying inside the radio beam when on its way towards the distant cluster. Additionally, the actual design of the message led to much debate. Is it really decodable for an alien intelligence? Perhaps it is influenced too much by human thinking. It may not be very obvious that the 1679 bits are supposed to be understood as an image where rows are formed from left to right and added to each other. The separation of data values and graphical information could also prove problematic. And perhaps the concept of "image" has a different meaning for an extraterrestrial intelligence, an ETI?

There are also technical objections to the transmission. It was never repeated, which otherwise is a typical requirement from SETI researchers in order to consider a signal genuine (see Chap. 11). Without such redundancy, it is enough to miss one single bit, compromising the idea of forming a 23×73 pixel image from the transmission. Also, the target of the signal is nowadays controversial. Scientists today consider it unlikely that star clusters like the M 13 are probable candidates for life-bearing planets, although we should keep in mind that the science of exoplanet research is very young. Finally, critics claim that the signal will not even reach the cluster since it will have moved out of the way once the signal arrives. These concerns, however, seem to be exaggerated.

The stellar cluster is very large, and modern values for its proper motion show that it is not moving faster than still allowing the signal to arrive well within its diameter.

Of course, it was not the intention of Frank Drake to begin an interstellar conversation. His message must be considered symbolic; that anyone will ever detect it is highly unlikely.

Modern Message Successors

The classical message from the Arecibo telescope has been followed by several more or less serious successors. These attempts are sometimes known as Active SETI (Search for Extraterrestrial Intelligence) or alternatively METI (where the M stands for Messages). A list of these can be found in Table 8.2. The Russian

TABLE 8.2 Some interstellar radio messages currently on their way to other stars

Message	Target	Constellation	Date of transmission	Date of reception
Arecibo message	Messier 13	Hercules	Nov 16,1974	Ca 27000
Cosmic Call 1	16 Cyg A	Cygnus	May 24, 1999	Nov 2069
	15 Sge	Sagitta	Jun 30, 1999	Feb 2057
	HD 17842	Sagitta	Jun 30, 1999	Oct 2067
	Gl 777	Cygnus	Jul 1, 1999	Apr 2051
Teen Age Message	HD 197076	Delphinus	Aug 29, 2001	Feb 2070
	47 UMa	Ursa Major	Sep 3, 2001	Jul 2047
	37 Gem	Gemini	Sep 3, 2001	Dec 2057
	HD 126053	Virgo	Sep 3, 2001	Jan 2059
	HD 76151	Hydra	Sep 4, 2001	Mar 2057
	HD 193664	Draco	Sep 4, 2001	Jan 2059
Cosmic Call 2	HIP 4872	Cassiopeia	Jul 6, 2003	Apr 2036
	HD 245409	Orion	Jul 6, 2003	Aug 2040
	55 Cnc	Cancer	Jul 6, 2003	Mar 2044
	HD 10307	Andromeda	Jul 6, 2003	Sept 2044
	47 UMa	Ursa Major	Jul 6, 2003	Mar 2049
Across the Universe	Pole star	Ursa Minor	Feb 4, 2008	Ca 2439
A Message from Earth	Gliese 581	Libra	Oct 9, 2008	Mar 2029
Hello from Earth	Gliese 581	Libra	Aug 28, 2009	Jan 2030

scientist Alexander Zaitsev has embraced the task of trying to communicate with ETI. He has been using a 70 m radio telescope in Yevpatorija in Crimea with a powerful transmitter. Relatively extensive messages were broadcast on four occasions—1999, 2001, 2003 and 2008. Contrary to the symbolic message of Drake, Zaitsev's transmissions have the direct purpose of finding another civilisation interested in communicating with us. They have targeted a number of nearby stars, most of them known to have planets. "Cosmic Call" transmissions were made in 1999 and 2003. The Zaitsev group then put together a longer message which took four hours to send, including some repeated sections. Part of the message included the "interstellar Rosetta stone", constructed by the Canadians Yvan Dutil and Stéphane Dumas. It constitutes a further development of the Arecibo message, but contains 23 binary images, each of the size 127×127. The idea is to use such pictograms to introduce a symbolic language that subsequently conveys information about Earth and humanity. Among other things it contains information about geometrical and mathematical relationships, atomic energy levels, the solar system, cells, etc.

Zaitsev sent the "Teen Age Message" in 2001, a dispatch using a slightly different approach. Zaitsev included one part containing musical excerpts, primarily classical music. Performers played the music on a theremin, an electronic instrument which produces well-defined sine waves, suitable for transmission as radio waves. The instrument, with its typical vibrato, gives a rather spooky impression. A well-known example of its use is the theme to the popular TV series "Midsomer Murders". The "Teen Age Message" also contained a digital portion where Russian teenagers could leave their own messages.

Communicating with the stars also has a considerable public relations value for commercial agents. A number of such messages have been sent, with dubious scientific value and even less (if possible) chance of ever obtaining a listener. "A Message From Earth" was transmitted in 2008, the result of a major internet-based project. Thousands of participants joined a competition which eventually resulted in the selection of 501 messages. These were put together with photo material in what was called a "digital time capsule" and transmitted towards the star Gliese 581. This star has at least three verified planets although there are disputed indications of several more. A similar project was "Hello From Earth" which was sent from Australia in 2009. This transmission was also directed at Gliese 581. A project called "Lone Signal", created a lot of publicity in 2013, claiming that it would allow the general public to participate in sending messages to the star system Gliese 586. However, in 2015 this project appears to have failed to achieve its goals.

Should We Send Messages at All?

The broadcasting of messages from Earth is far from uncontroversial. Despite good intentions, Zaitsev's activities have been the object of harsh criticism. Physicist Stephen Hawking have warned against too much interstellar openness. David Brin, scientist and well-known science fiction author, believes that it is irresponsible for a small group of people to send messages which, in principle, represent the entire human race. He argues that any attempts of contact must be preceded by careful discussions and consultations. It is not obvious that an alien civilisation would be harmless to us. Brin argues that there is no possibility to be sure of the motives of the aliens. It is undoubtedly true the discussion on contact activities suffers from a serious deficit of information, encouraging heavy speculation and badly founded conclusions. Perhaps we should follow the advice of Swedish philosopher Nick Bostrom and wait until we have a better basis for decision. However, signals transmitted cannot be taken back. Even if the risk appears very small, consequences could, in fact, affect the future destiny of humanity. We will later come back to that problem.

9. How Many Civilisations Are There?

Mankind has always been fascinated by the notion of other planets with potential life on them. But the question remains about if and where extraterrestrial life is to be found in the universe. The arguments for and against the existence of extraterrestrial life continue to cause lively debate. But at least today the subject can be discussed without fear for life and limb. In the sixteenth century things were far worse. The powerful Catholic church were suppressing new thought about the universe, Giordano Bruno and Galileo Galilei were among those punished. In more recent times, the Russian rocket scientist Konstantin Tsiolkovsky was among the first who presented modern thinking on this subject, in writings made in the 1930s but not discovered until far later, after the fall of the Soviet Union.

Is there any method that can give us a specific estimation of the number of planets containing life or even advanced civilisations? The astronomer Frank Drake achieved fame in the 1960s when he formulated the problem in a mathematical equation, which became known as Drake's equation. Originally the equation was not intended to be taken authoritatively or even literally. It actually stems from a meeting held in 1961 at the Green Bank radio observatory in the USA. Drake himself recalls:

"As I planned the meeting, I realised a few days ahead of time we needed an agenda. And so I wrote down all the things you needed to know to predict how hard it's going to be to detect extraterrestrial life. And looking at them it became pretty evident that if you multiplied all these together, you got a number, N, which is the number of detectable civilisations in our Galaxy. This was aimed at the radio search, and not to search for primordial or primitive life forms."

© Springer International Publishing Switzerland 2016
P. Linde, *The Hunt for Alien Life*, Astronomers' Universe,
DOI 10.1007/978-3-319-24118-0_9

Many people think that Drake's equation represents an optimistic view, since it presumes in some sense that other intelligent civilisations actually exist. However, as we will see, it depends very much on how the estimations are made. Both the optimist and the pessimist can use the equation, but we will start with estimations on the optimistic side.

The Drake equation requires the estimation of seven different factors which are multiplied together to form a result. Before we take a closer look at these factors we will briefly discuss a few more philosophical concepts which are useful as we penetrate the question of the existence of other possible civilisations.

Selection Effects and the Anthropic Principle

The term selection effect is actually derived from statistical analysis, but is often used in a more general context. The significance of the effect is that a conclusion runs the risk of being misleading if it is based on scant or erroneous data. A fundamental of all science it is to try to avoid—or at least compensate for—selection effects. A simple example relates to the fact that many big exoplanets have been found to be located close to their parent stars. This is a typical selection effect, since these planets simply are easier to detect. Still, it is possible that smaller, more Earth-like, planets may exist. Selection effects influence us persistently in daily life, for better or for worse, and it is wise to be aware of them, not least when you read your evening paper. Selection effects may play a major role in the discussion about ETI:s.

The philosophical concept known as the anthropic principle could be called the ultimate selection effect. In layman's terms, it states that things are the way they are because otherwise they could not have been observed in the way they are. One example of this is the natural laws of the universe, which contain a set of apparently fine-tuned constants (e.g. the gravitational constant). It is easy to show that our universe would have had a radically different appearance—with no possibility of life—if only one such constant had possessed a slightly different value. However, this fine-tuning is less remarkable if you adopt a more general cosmology, allowing for an infinite number of simultaneous universes. If you assume that the constants of nature in these other universes have different values, the conditions also would be drastically different. In such a case the values of the natural constants in our universe are not unique. However, for natural reasons (literally!) we live in that universe which, for us, has favourable living conditions. Proponents of the anthropic principle see a strong connection between existence of the universe and humanity.

Mediocrity Principle vs. Rare Earth Hypothesis

Another philosophical point of reference could be called the mediocrity principle, sometimes also known as Copernicus' principle. It could be described as a kind of cosmic equality law and is a reaction against man's constant tendency to place himself at the centre. Instead, Copernicus' principle asserts that the position of mankind in the universe is in no way exceptional. The Earth is not at the centre of the universe, the Sun is a most ordinary star at a normal location in a normal galaxy. This galaxy is only one out of billions of similar galaxies. So why would man as an intelligent being be so remarkable, or even unique?

The direct opposite of the mediocrity principle is the hypothesis of the rare Earth. It emphasises what is perceived as a set of unique circumstances, each of which a condition for the origin of life and its development on Earth. Proponents of this theory can be said to act as a kind of "devil's advocate" against the optimists. We will come back to this in the next chapter.

The Drake Equation

Now let us take a closer look at the Drake equation. It is usually formulated in the following way:

$$N = R \cdot f_p \cdot n_e \cdot f_l \cdot f_i \cdot f_c \cdot L.$$

N = the number of existing contactable civilisations
R_* = the number of stars created in the Milky Way per year

f_p = the fraction of stars that has planets

n_e = the number of habitable planets per star with planets

f_l = the fraction of habitable planets that develops life

f_i = the fraction of life-bearing planets that develops intelligent life

f_c = the fraction of planets with intelligent life that develops a technological civilisation that is capable of interstellar communication

L = the life span of a technological civilisation

Let us examine these factors in some detail.

How Many Stars Are Created in the Milky Way Each Year?

Initially it may seem entirely impossible to estimate the number of stars created each year in our galaxy, the Milky Way. The stars in the sky seem to be unchangeable to most people; no stars seem to be added or subtracted. Nevertheless, we know that new stars are constantly being created from clouds of gas and dust. We can observe the continued activity of this process everywhere in the Milky Way using telescopes. The spiral structure of the Milky Way is in itself a strong indicator of star formation. The spiral arms are actually not physically connected structures, but are due to a wave phenomenon which effects the whole galaxy. Interstellar gas is compressed by the wave, stimulating star formation. In turn this means that star formation, including bright stars, is enhanced in such locations. The space between the spiral arms is not empty but is dominated by fainter and older stars. Consequently, the arms of a spiral galaxy do not follow the same rotation as the stars of the galaxy do. At the Sun's distance from the centre of the Milky Way it takes about 225 million years for a single orbit around the centre. This is known as a galactic year.

For stars a strong correlation exists between properties such as brightness, age, colour and mass (see also Box. 6.4). Without going too far into astrophysics, we can conclude that stars are created many at a time from giant gas clouds, nebulae, that are dominated by hydrogen gas but also contain lots of minerals in the form of dust particles. The termination of the star formation from a cloud results in the creation of a stellar cluster. These stars have varying masses. A few have a mass much larger than the Sun's. They are called blue giants and are very hot and luminous. In spite of their large initial mass they consume their available energy at a very fast rate. After just a few hundred million years, most have already burned out (and may perhaps form black holes). Stars like the Sun are much more long-lived, the red dwarfs even more so. Red dwarf stars are the most common ones. They may have a tenth of the mass of the Sun, but in return they deplete their energy at such a low rate that they can exist for hundreds of billions of years.

So, the larger the mass, the hotter and bluer and brighter and more short-lived the star. The lesser mass, the cooler and redder and fainter and more long-lived.

Fig. 9.1 A typical star formation field. The red gas is hydrogen gas, the dark areas contain dust. A number of blue giants have already been born (ESO)

In our local surroundings in the Milky Way, let us say within the nearest 10,000 light years, we see lots of evidence of star formation in the form of nebulae with newly created stars (see Fig. 9.1). The most well known example is the Orion Nebula. This is visible with naked eye as a small diffuse patch, and is our closest star formation area, at a distance of about 1350 light years. Nebulae of this type are not small. The Orion nebula, for instance, has an extent of 24 light years.

We know that the oldest parts of the Milky Way are more than 10 billion years old. At the same time the number of stars in the Galaxy is estimated to be about 300 billion. If the star formation had been constant and all created stars still remained (which we know is not the case) you could get a rough estimate by dividing these numbers with each other. The result would be 30 new stars per year. However, it seems reasonable that star formation should have been more intense when the Galaxy was younger than now, and one could ask whether not all gas already has formed stars. However, this is not the case. The younger (and heavier) generation of stars often eject part of their mass before they die. This matter disappears out into the interstellar medium, but may be recollected to create "kindergartens" for a new stellar generation. It is interesting to notice that this matter is enriched with heavier elements that were created in the nuclear processes inside the dying stars. These elements are fundamental for the creation of Earth-like planets. In addition to this recycled matter, there are also observations that seem to indicate that the Milky Way is still receiving gas streams from the intergalactic medium.

Some efforts have been made to directly measure the star formation rate in the Milky Way, among others with the help of satellite observations. The result from these studies is that on average, seven stars are created annually. This is the value that we will adopt.

How Many Stars Have Planets?

When Drake formulated his equation, the existence of planets around other stars was still just speculation. But of all the factors that are part of the Drake equation, this factor is the one where we now have new information with some degree of reliability. As is seen from Chap. 6, since 1995 we have experienced an increasing stream of exoplanet discoveries. However, this research field is still in its infancy, and it is natural that the current discoveries primarily represent those planets that are easiest to discover.

So far, a more general estimation of the number of planets must still be based on statistical arguments, albeit with the difference that now there is a firm basis from which to proceed. In addition to all new information about exoplanets, we have seen a lot of news about the stars themselves. It has become clear that a large fraction of all stars are double or multiple systems.

Double stars exist in many types. We can distinguish well-separated double stars in small telescopes. A beautiful example is Albireo in the constellation of Cygnus, which contains two stars with conspicuous colours. One is red-yellow, while the other is blue. The stars are assumed to orbit each other with an orbital period of about 75,000 years and at a distance of 4000 AU. This distance is about 100 times larger than the diameter of our solar system, and there is probably nothing that precludes any of the components in having planetary systems. However, many other double stars are much closer to each other. Some are even in physical contact with each other, and continuously exchange mass. These stars have orbital periods of only days or even hours. It seems very unlikely that such systems would have planets in sufficiently stable orbits to be of any interest to us.

On the basis of available information we suggest, perhaps somewhat optimistically, that half of all stars have planetary systems.

How Many Habitable Planets Are There Per Star?

We return momentarily to the concept of a habitable planet. In fact one might consider the meaning of both the word "habitable" and the word "planet". Let us start with the latter. Until the turn of the millennium it was rather straightforward to define a planet. Since 1930 we traditionally had nine main planets in the solar system. But then something happened to complicate things, and the astronomers struggled to bring some order. In the beginning of the twenty-first century, as we have seen, astronomers discovered several new planet-like objects in our own solar system. The new solar system objects turned out to belong to a newly found belt of remnants from the childhood of the solar system, known as the Kuiper belt. This zone of debris extends far out in the solar system (see Chap. 2). Although the newly found objects were large, they resembled more the asteroids that lie in a belt between the planets Mars and Jupiter. The increasing confusion led to an intense debate among astronomers, eventually leading to a new definition of the term planet. The modern definition proclaims that (a) a planet must be orbiting the Sun, (b) have enough mass and gravitation to form a round shape and (c) dominate its surroundings so that it is free of similar objects.

Defining the concept a habitable planet is certainly also not straightforward. Habitable for what and for whom? The most common definition refers to the properties of water. If you assume that the emergence and development of life requires liquid water, this immediately limits rather clearly the allowed temperature range for a planet, and thus its distance to its star. It should neither be so warm that the water permanently boils away nor so cold that it remains ice forever. So we call this zone the habitable zone or the ecosphere. So far, we only have one planetary system to study at close range, our own. The new information from the numerous exoplanets is, after all, still very incomplete, and anything beyond rather vague conclusions from statistical data is difficult to establish.

The size of the ecosphere of the Sun is not undisputed, as we saw in Chap. 3. Within the ecosphere we have the planets Venus, Earth and Mars, and in fact marginally also the asteroid Ceres. The Moon also qualifies. Other scientists claim that the habitable zone is considerably narrower. As we have seen, Venus lies so close to the Sun that a catastrophic greenhouse effect has caused an avalanche increase in temperature to several hundred degrees. Likewise, we have concluded that Mars most probably looked very different at a very early stage. Large amounts of liquid water must have been present, and the atmosphere must have been denser than now. Clearly, conditions for emerging life were much better than today, and the question of whether there was—or is—life on Mars remains to be solved.

Another non-negligible factor for the ecosphere is that it must also be constant in time, preferably over a few billion years. Scientists know that the luminosity of solar-type stars varies with time, with the consequence that the habitable zone moves and changes in size over a longer time perspective. On the other hand, we know that our own Sun has changed in luminosity since the early days of the solar system formation, and life on Earth has nevertheless survived, partly because of the temperature regulator which is inherent in the climatic mechanisms of Earth (see Fig. 3.3, Chap. 3).

So it is a necessary but not sufficient condition that a planet is located inside the ecosphere. Or, is it perhaps not even necessary? We have seen in Chap. 4 that life might exist in rather unexpected and unusual places in our solar system. Several very large moons orbit around their host planets. Two of them are larger than Mercury and another three are almost as large. Since Pioneer and Voyager took the sensational close-up pictures of the satellite system of Jupiter in the 1970s, it has been recognised that suitable conditions for life may exist on at least one of the Jupiter moons, Europa. In spite of its long distance from the Sun it is still heated, but instead from tidal effects of the giant Jupiter. How many moons of this kind exist in other planetary systems is anybody's guess, but there could be many. Something we already have learned from exoplanet research is that many planetary systems look very different from ours. Inhabitable Jupiter-type giant planets often orbit rather close to their parent star. But there is no reason to believe that they cannot have large moons that are more similar to Earth.

With this slightly generalised definition of a world suitable for life, what would then be a reasonable guess for number of habitable planets per star with a surrounding planetary system? Let us choose the number 3.

The first three factors of the Drake equation that we have so far discussed are all in the realm possible observations, and recent advances have certainly added to a sounder basis for estimations. But the four remaining factors suffer strongly from information deficit and are of a much more speculative nature. Since the estimation for these factors vary widely, so will the end result for the equation. The recurring problem is that we have only a single example to refer to for these factors, namely our own. Keeping the anthropic principle in mind, one must at the same time realise that statistics based on a single example are more or less worthless. But it is the only thing we can extrapolate from, so let us boldly look at the remaining factors and try to estimate them.

How Probable Is the Emergence of Life?

Let us assume that we have a planet with suitable conditions for life—how likely is it that life emerges spontaneously? So far, our estimations based on observations and established facts. But here we immediately tread on thin ice. With very few facts available, we must fall back on speculation and sometimes even pure guesswork.

Even in the case of the Earth, we cannot with any certainty explain why and how life was created. Some researchers avoid this fundamental problem by arguing that life has arrived on Earth from outside (see Fig. 9.2). The Swedish physicist and Nobel laureate, Svante Arrhenius, was one of the advocates of this theory in the beginning of the last century. His idea was then that a kind of life spore, panspermia, could survive during long periods of time inside space-faring small bodies like meteoroids, asteroids and

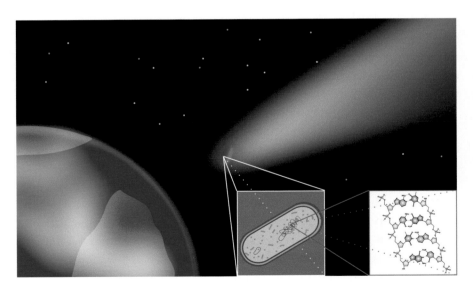

FIG. 9.2 According to the panspermia hypothesis life has arrived to Earth via colliding bodies (Silver Spoon Sokpop/Wikipedia)

comets. Even if the theory does not answer the ultimate question of the actual origin of life, it at least means that life should be common if other life conditions are good. Theorists have put forth several modern variants of the theory.

In the 1980s, the famous cosmologist Fred Hoyle advanced the theory that primitive life would form and exist in the innards of comets, where pockets of liquid water could exist. He proposed that life had come to Earth via comet collisions, and even believed that the recurrent virus diseases were connected with comets. As late as 2013 British scientist Milton Wainbright claimed they had found evidence for extraterrestrial life in the stratosphere at a height of 25 km.

The Murchison Meteorite

On September 28, 1969 inhabitants of Murchison in Australia witnessed a bright flash in the sky. It was a meteorite which split into smaller fragments on its way down. The impact site was identified and the total sum of the collected bits weighed more than 100 kg.

The meteorite turned out to be of the carbonaceous chondrite type. Experts assume these meteorites are very old, containing the kind of original material that once formed the solar system. Carbonaceous chondrites can be considered as a kind interplanetary fossil. With time the Murchison meteorite has become one of the most well-studied. Researchers estimate that the meteorite is actually old enough to have been moving around in the original cloud from which our solar system formed. Inside the meteorite, a large amount of complicated organic molecules could be identified. An advanced analysis made in 2010 showed traces of 14,000 organic compounds and 70 amino acids. One should however be careful when interpreting results from meteorites. After creation they may have been exposed to various environmental effects in space. And after landing on Earth they may be contaminated by terrestrial chemistry. Today, very few would argue that life emerged in the original planetary nebula, long before Earth had formed. But it is nevertheless evident that many of the complicated molecules that are needed for life actually spontaneously form in "empty" space.

How and if life was created on Earth by itself is a fundamental question that still awaits an answer. As we have seen in Chap. 3, the first traces of life date to about 3.5 billion years back, less than 1 billion years younger than Earth itself. We should also consider that the young Earth was heavily bombarded by asteroids and meteorites that probably kept the surface hot and liquid. The very fact that life emerged so soon after reasonable life conditions were at hand suggests that life actually is created relatively smoothly and easily.

Biochemists have made many attempts to create life artificially in order to better understand the origins of life. One such classical experiment was carried out in 1952 by Stanley Miller. In a laboratory setup, Miller attempted to reproduce the assumed atmosphere of primordial Earth. A mixture of hydrogen, methane and ammonia came into contact with water and was subjected to electric discharges. After a week, up to 15 % of the carbon in the system had reorganised itself into a number of organic compounds, among them more than 20 amino acids. If these experiments had been more successful, then the puzzle of the origin of life would come closer to a solution.

But the step towards creating even more complex molecules, e. g. proteins and RNA, is very large, not to mention the step to the most defining qualities of life: replication and metabolism. A number of mechanisms have been proposed, but it would lead too far outside of our context to go into them in detail. Some scientists are of the opinion that the emergence of life on Earth is the result of a long series of improbable events and conditions. Others, however, say that life may very well have been created several times on Earth, independent of each other, but no evidence of this has ever been found.

Whether life is formed just because it can be formed is still an unanswered question. If investigations can confirm that life has independently emerged in other places in the solar system, e. g. on Mars, it would be a revolutionary discovery. Then we could, with a clear conscience, assign a value of 1 to this factor.

So where do we end up in our estimation of how often life emerges given the right conditions? We will have to make a guess, probably on the optimistic side, when we follow Drake's original suggestion, namely 0.5.

Does Life Always Evolve to Intelligence?

Lately, the matter of the development of life after its beginning has been the topic of an intensive debate. The growing realisation concerning the irregular advancements of life, evidently quite often influenced by purely random effects, has caused many to doubt the foregone conclusion that primitive life always evolves towards intelligence.

We note that it has taken a very long time for nature to create intelligence at a technological level on Earth. After about 3.5 billion years, only one single species has achieved the capability of a space exploration programme and of interstellar communication. Only within about the last 100,000 years have we seen undisputed intelligence, and only within the last 100 years have the means for remote communication been available. We do not get the impression of a slow, steady and unavoidable evolution, but rather the implication that coincidences play an important role. Why has it taken so long? Is intelligence really an indispensable survival factor? Or is it, perhaps, the other way around; that the intelligence evolution on Earth has been unusually slow in progress? And, for that matter, what do we actually mean by intelligence? None of these questions are very easy to answer.

Among the most distinguishing marks of intelligence, we normally point at characteristics like consciousness, time perception, capacity for effective communication, capability of overall analysis and curiosity. Is the existence of these qualities in mankind just a coincidence; is it an ecological niche that must sooner or later be filled by evolution?

Yet again we have a situation where the only example we can point to is Earth. The best one can do is to study in detail the evolution of life here and try to draw conclusions. For example, is it possible to see obviously independent evolution that has resulted in the same type of properties, perhaps even the same appearance? Finding distinctly different evolution paths on Earth is for natural reasons not easy but it is not entirely futile.

Life on Earth displays many examples of how creatures with different background and development nevertheless are quite similar to each other. Critical organs, as for example eyes, have the same advanced function among both vertebrates and cephalopods (such as squid, octopus, etc). The actual growth of intelligence is harder to get a grip on, but we know that aquatic mammals, especially dolphins, also exhibit several of the qualities that we associate with intelligence. One interesting factor for the development of a creature's intelligence seems to be the duration of its gestation period and rearing. Human children are unusual, although not unique, in this respect, since they need protection and education for a considerable time. It could be described as man not being very "hard coded" but instead more "software programmed". But apes and marine mammals also show more or less the same behaviour. It is difficult to speculate on what would happen to the semi-conscious animals of today if mankind disappeared from the scene, but it is not hard to imagine that some other species, given enough time, could evolve in a way similar to mankind. There is no doubt that man has clear advantages in his anatomy, allowing for mobility on land and a manipulative capability with hands and fingers, developed from the apes' original necessity to swiftly move among trees.

Judging from our only example from Earth, even if we may have some grounds for optimism when it comes to the evolution from single-cell life to intelligence, there is little basis for applying this to other planets. We simply must concede our ignorance on this matter. Nevertheless, let us optimistically assign the factor f_i with a value of 0.1. This thus implies that on every tenth world where life has emerged, life will evolve into intelligent creatures.

Are All Civilisations Interested in Technology?

The evolutionary direction of an alien civilisation is very hard to hypothesise, and again we are reduced to little more than pure speculation. As usual, we have more questions than answers. First we must try to define what we mean by civilisation and what we mean by technology, and this gets us into trouble already. A civilisation in earthly terms involves, among other things, organised food production, city establishments, existence of writing and mathematics as well as economical and political centralisation. An organised religion is also often ascribed to civilisation. From the standpoint of such a definition, many civilisations have existed on Earth (and disappeared). But to apply these concepts on other worlds presumes Earth-like conditions. Fortunately, the definition for technology in this context is easier. Our somewhat restricted definition must involve capacity and interest in communication at interstellar distances. Our current civilisation has had, in a limited sense, this ability since the beginning of radio transmissions. The driving force behind this achievement is a constant curiosity and an urge to improve and simplify our lives. Ever since the fire and the wheel, technological development has been evolving toward a single goal, to improve our standards of living. Further, this has happened at an incredible speed during the past century. Furthermore, the need for communication has strongly stimulated the development of radio and television technology. At the same time we should note that standard of living and quality of living are increasingly considered as separate things. Perhaps our technological evolution is not necessarily continuing with dizzying speed? We will return to this discussion in Chap. 13.

It is only with considerable difficulty we can bend our fantasy far enough beyond our everyday limitations to envision totally alien civilisations. Authors in the science fiction genre have often made exciting attempts. From a more scientific point of view, we must admit that other habitable worlds may have very different conditions than those on the Earth. We have reasons believe that water worlds may be quite common, i.e. planets that are completely covered by water, not just to 70 % as on Earth. In such cases civilisations must evolve under entirely aquatic conditions. It is not obvious that discoveries like the fire and inventions like the wheel have any meaning on such worlds. Perhaps a highly advanced aquatic civilisation would never conceive of the existence of an outside universe.

Yet, again we chose a somewhat optimistic estimation of f_c, i.e. the fraction of civilisations that evolve a technology for interstellar communication. We assign it the value 0.5.

What Is a Normal Survival Time for a Civilisation?

Clearly, the most difficult task of all is to estimate the lifetime of a technological civilisation. Unfortunately, as we will soon see, it is also the most decisive factor. The only available information we have is that our own civilisation has survived so far for almost a hundred years. But during this time period the technical development has been staggering. Even an attempt to predict what our world will look like in another hundred years, not to say a thousand or ten thousand years, appears almost hopeless. We have many examples of earthly civilisations (or rather cultures or empires) that have had limited longevity; the Roman Empire, the Mongol Empire and the Maya culture are some examples. Their longevity can be measured in a few centuries. But none of these examples are very illuminating since they have not been anywhere near the technological level of our current civilisation. Nor have they been global civilisations with efficient communications.

Others point in a more general way to the example of the dinosaurs, which dominated Earth for nearly 150 million years. However, this is a rather meaningless comparison. Still others note that from prehistoric studies we may conclude that the typical life span for a vertebrate species is 2–4 million years. But humanity can understand and affect its future destiny in a way that has not been possible before. For instance, no other vertebrates have been capable of understanding the threats to their existence.

Our civilisation has reached a level, almost at a crossroads, where problems must be dealt with on a really global scale. As we know, for the past few decades humanity has had the capacity to destroy itself in a nuclear war. This threat has de facto diminished during the latest decade, which is encouraging, but it needs to be entirely eliminated. The population explosion, with its accelerating demand for resources, is another difficult problem awaiting a solution. Additionally, these problems must be solved in harmony with environmental considerations in order to avoid drastic climatic changes. Is humanity capable of solving these problems? The answer will become apparent during the next hundred years, after which it is most likely too late. In the next chapter we will look closer at the survival prospects for a civilisation.

So where do we wind up in our estimation for the average life span of a technological civilisation? Let us, with many reservations, suggest 50,000 years. This guesstimate at least has the advantage that some would consider it pessimistic, while others would think it overly optimistic.

So How Many Civilisations Are There?

We have now made an estimation for all the factors in the Drake equation. Let us see the result we have derived:

$$N = R \cdot f_p \cdot n_e \cdot f_l \cdot f_i \cdot f_c \cdot L => N = 7 \cdot 0.5 \cdot 3 \cdot 0.5 \cdot 0.1 \cdot 0.5 \cdot 50{,}000 = 13{,}125.$$

According to this, more than 10,000 civilisations should exist in the Galaxy right now! It seems extremely encouraging (although others would consider this as threatening). From the formula, the importance of the dubious estimation of a civilisation survival time is obvious. If we replace this value instead with 1000 years, the answer would be 263. But the pessimists will have their say in the next chapter. Let us conclude this chapter by making an estimation of the average distance between two civilisations in the optimistic case. Using a simplified approach we assume that they are equally spaced throughout the Galaxy. The diameter of the Galaxy is around 100,000 light years, and its average thickness is about 1000 light years. This gives a volume of nearly 10 trillion cubic light years! In consequence, this results in an average distance of about 900 light years between two technologically advanced civilisations. And this is a long distance.

10. Where Is Everybody?

In the summer of 1950, the famous Italian physicist Enrico Fermi worked at the Los Alamos National Laboratory in New Mexico, USA. Fermi had, in 1938, received the Nobel prize for his pioneering discoveries concerning radioactivity and nuclear reactions. He was somewhat of a universal genius, unusual in his ability both for theoretical thinking and for practical experimentation. Now he was working as a consultant investigating the possibilities of building a hydrogen bomb, together with a group of other world-famous physicists.

During a lunch an event transpired which has subsequently become famous. The conversation was easy-going. The New Yorker Magazine had just published a cartoon blaming the disappearance of municipal trash cans on marauding aliens; An invasion of little green men in flying saucers was responsible. Fermi jokingly commented that it was a good theory since it explained both sightings of flying saucers and the disappearance of thrash cans. The discussion continued on the topic of whether flying saucers could exceed the speed of light in some way. The conversation then moved on to more mundane matters. Suddenly Fermi, who had apparently been mulling the previous subject, exclaimed: "But where is everyone?" Everybody present understood what he meant. And with that the Fermi paradox was formulated.

Fermi (see Fig. 10.1) was a master of making rough estimates of the most curious phenomena. He encouraged his students to use their own thinking and experiences to arrive at meaningful, albeit approximate answers. A classical Fermi question is "How many piano tuners are there in Chicago?" (The answer was, after a number of estimations, about 100). At the lunch in Los Alamos he calculated how many civilisations should exist in the Milky Way, using principles that would later be clarified by the Drake equation. The result was the reason for his exclamation. There should be a lot of them. Consequently, he said, should we not already

© Springer International Publishing Switzerland 2016
P. Linde, *The Hunt for Alien Life*, Astronomers' Universe,
DOI 10.1007/978-3-319-24118-0_10

FIG. 10.1 Enrico Fermi in about 1950 (NARA)

have seen traces from all these aliens? Since this is not the case, Fermi commented, thus the paradox. A more extensive definition of the Fermi paradox can be stated in the following way:

The apparent size and age of the universe suggest that many technologically advanced extraterrestrial civilizations ought to exist. However, this hypothesis seems inconsistent with the lack of observational evidence to support it.

Enrico Fermi's question has been hanging in the air ever since, and many plausible explanations and speculations have been put forward to provide an answer. We will now take a closer look at some of them.

Colonisation of the Milky Way

The core of the Fermi paradox is the proposition that our galaxy, the Milky Way, should have been completely colonised for a long time. Long ago we should have discovered life in some way or other, either directly or indirectly. Many simulations have been made in order to model and understand a hypothetical galactic colonisation. The simulations take into account the physical laws as we know them, in particular the fact that the speed of light constitutes an upper limit for how fast it is possible to travel. But it is also assumed that interstellar space travel is indeed possible, even though it can be very time-consuming. Of course, an important aspect is what motive an alien civilisation would have to colonise our galaxy. Since we can hardly acquaint ourselves with the thinking processes of an alien, we have to settle for experiences from our own world. We then consider some possible driving forces. Hard pressures from dwindling resources due to a rapidly growing population could be one such factor. Scientific curiosity might be another. Still other motives from the history of mankind include escape from war or persecution, or simply a desire to secure the survival of a culture, a people or an entire race.

Depending on the initial values of the models for a colonisation, scientists have arrived at different conclusions. Let us begin with a terrestrial example as a model, namely the colonisation of North America. There the population grew by 3 percent a year, and the immigration rate was 0.3 per mille a year. Translated into galactic conditions, a corresponding process would fill the entire galaxy in 5 million years. Every habitable planet should already be colonised!

Some scientists claim that population increase plays no role at all. Humanity has tripled its population during the last 70 years. Should this trend continue during the next centuries, we would soon need to colonise new worlds, literally speaking, at the speed of light. Carl Sagan suggested that the population explosion problem clearly needed to be solved on site, independent of interstellar colonisation. If we again use Earth as an example, we can note that the rate of increase in population actually has decreased, and there is some hope that the population on Earth can stabilise around 10–15 billion people.

Geoffrey Landis has put forward an interesting model for the colonisation of the galaxy. Landis works with future space projects for NASA and is also known as a science fiction writer. He analyses the colonisation process using a mathematical tool called percolation. In simplified terms, the process can be compared to a coffee-making machine, where water randomly and gradually penetrates the ground coffee. In this comparison the gradually flowing water is the civilisation process while the coffee grains correspond to the stars in the Milky Way.

Landis starts out with three assumptions and one rule. The first assumption is that interstellar travel is possible, but expensive, slow and impractical. Therefore, a limited zone exists within which colonisation is practical. Perhaps the radius of such a zone could be 50 light years. The second assumption is that the long distances between colonies unavoidably allows for only a very weak central control. With time, this means that every new colony develops its own culture and needs, and in practice becomes independent of the home world. The third assumption is simply that a civilisation cannot colonise an already colonised world. This is partly a consequence of the first assumption—if interstellar journeys are difficult and expensive, it must be even more expensive and difficult to mount an invasion.

Landis also introduces a simple rule: Either a culture has the driving force to continue its colonisation or it does not. If the driving force is lacking, it may be explained by the fact that they are incapable of colonising (because the nearest star system is to far away), or simply because they do not want to do it.

The percolation problem is simple to simulate in a computer and provides concrete and interesting results. In Fig. 10.2 the result of such a simulation is shown. The black circles correspond to civilisations that colonised their immediate surroundings, the grey circles correspond to civilisations that have stopped to colonise and the white areas are unexplored. The end result of the simulation depends strongly on the probabilities that control Landis' rule, i.e. the capability and will to continue colonisation. Zones will always exist that are never colonised.

Landis' percolation description has the advantage of being more realistic than the majority of other theories. It also explains why an ongoing colonisation not yet reached Earth, nor will it ever. We simply are located in a white unexplored zone. This could be a solution to the Fermi paradox. But Landis' assumptions can also be criticised. For example, how likely is it that there really exists a "distance horizon" beyond which it is not possible to continue expansion? Also, Landis' scenario implies that it is the aliens themselves that are doing the colonisation. But perhaps they only dispatch machines? And as such they are perhaps not subject to the same limitations.

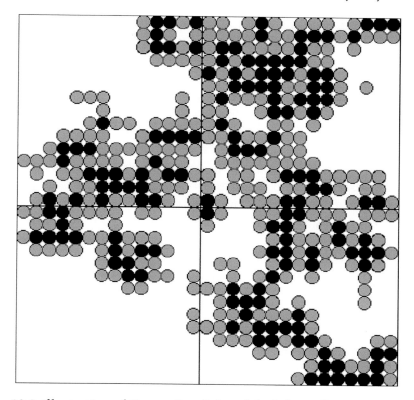

Fɪɢ. **10.2** Illustration of George Landis' model of the Milky Way colonisation. Dark circles correspond to colonies who want to continue colonisation. *Grey circles* are colonies who have stopped colonising, while white areas are unexplored (Peter Linde)

Von Neumann Machines Conquering the Galaxy

In discussions involving the possible existence of alien civilisations, the von Neumann machines usually play an important role. World-famous mathematician John von Neumann made important contributions in a variety of fields in mathematics, as well as laying the ground for the development of computers.

He also developed ideas of self replicating machines, i.e. advanced machines capable of reproducing themselves in multiple copies, assuming that basic resources were within reach.

Mankind has already sent interplanetary probes to most parts of the solar system, and a few are in the process of leaving it.

Hence, the very idea of sending interstellar probes in a similar way is not too far-fetched. Intelligent von Neumann machines could be dispatched with the aim to explore, identify resources and even colonise neighbouring worlds. After landing on an alien world (i.e. an exoplanet), such machines set to work making copies of themselves, and then would launch these to continue the exploration towards new stars.

Frank Tipler, an American mathematical physicist, further developed this reasoning. In 1981, he postulated that a robot of this type could achieve a speed corresponding to 10 % of the speed of light. If we further assume that the average distance between two stars in the Milky Way is five light years, such a trip would require 50 years. Assuming that 100 years are needed in order to explore the planet, create copies of itself and start all over again, it is rather easy to calculate that even at this moderately slow speed our whole galaxy would be thoroughly explored in about 10 million years. If you are satisfied with a maximum speed of only 1 % of light speed, the same result is achieved in 100 million years.

The notable thing is that both of these periods of time really are quite short. Our solar system is almost 5 billion years old, the Milky Way is about twice that age, i.e. 10,000 million years. Surely at least one single civilisation ought to have carried out this colonisation? But still we see no traces of it. This is sometimes called "the Great Silence". Something must have gone wrong. Or perhaps the logic of the reasoning is too simplified or has a serious inherent flaw? We need to look closer.

Bracewell Probes

Let us consider a somewhat more investigative and less aggressive version of the von Neumann self-replicating robots. In 1960, Ronald Bracewell, professor at the Stanford University in California, USA, suggested a more sophisticated variety of the von Neumann machines. His type of robot, a Bracewell probe, would not just make copies of itself, it would also systemically explore an alien planetary system in order to subsequently lay the ground for future colonisation. It would also carry embryos which could be awakened when needed. The Bracewell probe thus would be capable of a more directed and planned colonisation.

As indicated previously, a Bracewell probe would not be limited by any kind of "distance horizon" as in Landis' scenario. Nonetheless the question remains as to why we have not yet encountered this kind of powerful probe, which by itself would have the capacity to reorganise a planetary system. But perhaps there are more discrete and observing probes, perhaps even in our immediate vicinity. Again, we have no evidence for this. On the other hand, if such a probe existed inside our solar system, and did not want to be discovered, we would probably find it exceedingly difficult to find it. But we cannot entirely rule out the possibility that an alien civilisation has placed surveillance equipment in our surroundings. The excellent science fiction movie "2001—A Space Odyssey" put forth exactly this theme. The author, Arthur C. Clarke, speculated about an extraterrestrial intelligence which interfered in the evolution of Earth by mentally stimulating our distant ancestors. In the movie, in 2001, astronauts on the Moon finds a hidden monolith which transmits a message back to its creators at the moment it is struck by sunlight for the first time after millions of years.

A Bracewell probe certainly would have little difficulty in keeping itself hidden. The Moon offers good opportunities, and if one wanted to keep a distance, there are many hiding places out in the asteroid belt. Some searches have already been undertaken, for example in the so called Lagrange points. These are places out in space which provide stable parking places for a space probe. The gravitation from two nearby major celestial bodies can interact in such a way that there are two points in the neighbourhood that are stable against disturbances. Both the Earth-Moon system and the Earth-Sun systems have such places. Lagrange points could be ideal observing points for Bracewell probes. However, no alien object has so far been identified. Meanwhile, terrestrial research satellites are already occupying such places.

One suspected case does exist. In 1991 a small object passed very close to Earth. It was discovered by Spacewatch, a project at the Kitt Peak observatory in Arizona, USA, which has monitored the sky since the 1980s. Every dark night, astronomers there search for asteroids with mid-sized telescopes using advanced automatic cameras. Now and then they find objects in the proximity of Earth. 1991 VG turned out to be a very unusual object, since it had an orbit around the Sun that was very similar to Earth's orbit. The object showed very rapid light variations which were interpreted, by some, as reflections from a rotating metallic vessel. The size was estimated to about 10 m. Observers quickly checked to see whether the object could be explained by some old rocket stage—for example a leftover from the Apollo programme—which might exhibit such behaviour. But this solution was impossible to confirm. Then it was suggested that the object could be a Bracewell probe dispatched from an extra-terrestrial civilisation. But this could not be confirmed either. Later analysis and new discoveries have shown that 1991 VG instead may originate from a new type of very small asteroids, so called Earth trojans. Perhaps a more conclusive answer can be obtained in August 2017 when it will pass close to Earth once again.

Other observations have been connected to the possibility of an alien spacecraft. Beginning in 1927 there have been several observations of strange radio echoes, with a delay of several seconds. The phenomenon is known as Long Delayed Echoes (LDE) and involves the return of a radio transmission as a weak echo. Echoes longer than 2.7 seconds are difficult to explain since this is the time needed for radio waves to bounce off the Moon. However, longer delay times than that have been seen and must be due to an even longer travelling distance. Could it be an intelligent probe manifesting itself by reflecting transmissions from Earth? The origin of the phenomenon is still unknown, but prevailing theories explain it either as reflection against plasma clouds out in space caused by solar eruptions or as multiple reflections within the ionosphere and magnetosphere of Earth.

Could a Bracewell probe ever be created on Earth, or for that matter, in any other place? The answer is, perhaps somewhat surprising, that something similar already exists. Nature did it on Earth several billion years ago. Interesting enough, a cell is a biological automaton: it has several fundamental properties in common with a von Neumann machine. It can clearly replicate itself, and it incorporates extensive programming, in the form of DNA. Whether it is in the process of colonising the Milky Way remains to be seen…

So which plausible explanations can we find for the Fermi paradox? As in other information-starved discussions, we have to resort to speculations. There are undoubtedly several to chose from, many of them both interesting and entertaining. We are going to inspect a few of them. They can be roughly divided into two types. One type contains theories that include the existence of aliens and tries to explain why they have not yet been observed. In this group we find both optimists and pessimists, and of course all scientists that are actively searching for ETI. The other main group simply assumes that aliens do not exist and tries to explain this fact. We start with the latter.

ETI Do Not Exist—We Are Exceptional

In 2000 Peter Ward and Donald Brownlee published a book with the title *Rare Earth: Why Complex Life Is Uncommon in the Universe*. In it, they listed a number of arguments in favour of a theory which claims that life on Earth really is the result of an almost unique combination of several random factors. In a sense, their position was a complete antithesis to mediocrity principle, which states that the Earth and humanity are in no way special or extraordinary. The expert background of Ward and Brownlee as geologist and astronomer, respectively, gave the arguments a sufficient weight to provoke a discussion which is still going on. So, how unique are Earth and our solar system? Many of the arguments concern the difficulties for life to emerge and thrive, something that we already touched upon in the chapter about the creation of life. We now return to investigate these factors in more detail.

The Sun—Not Your Average Star

Starting by looking at the Sun and its location in the Milky Way, we find that our star lies in a relatively calm and protected zone, at a distance of about 30,000 light years from the centre of the galaxy. We now know that this galactic centre (as is the case for many other galaxies) is dominated by a fantastic phenomenon, a giant black hole (see Box 10.1), with an estimated mass of about four million solar masses. In the proximity of the black hole we observe high levels of gamma and X-ray radiation. This condition could rule out the possibility of life emerging near the centre of the Milky Way (Fig. 10.3).

Box 10.1: The Black Holes

Modern theories for stellar evolution offer predictions for what happens when the star exhausts its nuclear fuel. This happens at different rates for different stars; the fuel is consumed much faster for heavier stars than for lighter ones. In fact, heavier stars have a shorter life span. When the nuclear fuel has expired at the stellar centre, the rest of the stellar matter collapses, shrinking very quickly. Smaller stars, such as our Sun, end up as white dwarfs, with a size similar to Earth. Larger stars end up as rapidly rotating neutron stars, not larger than about 20 km! The density in these collapsed objects is enormous.

A black hole is the third possible end point. Black holes may result from larger stars, which still have at least 3–4 solar masses remaining after they violently shred outer layers in—for example—a supernova explosion. In this case, no counter-acting force is strong enough to stop the collapse of the stellar matter. The density and the gravitational forces become extreme, creating remarkable effects on the immediate surroundings. Light itself can no longer exit the object, hence the term "black". Additionally, space and time are affected close to and inside the object. Space-time becomes strongly warped, which justifies the term "hole". Even if the black hole cannot directly be observed, infalling accelerating matter emits intensive radiation which can be observed. Larger black holes which have swept up their surroundings can still be observed through their gravitational effect on the motion of nearby stars.

There are at least two types of black holes. Some form from single stars. But since nothing stops a black hole from growing unrestrained, there are also black holes that have consumed millions of stars. This type of super-massive black hole is often found at the centres of galaxies, including our own Milky Way.

Astronomers speculate about the existence of a third type of microscopic black hole, which in theory could have been created during the Big Bang. So far, no evidence of this has been found.

FIG. 10.3 An artists vision of a super-heavy black hole (NASA/JPL-Caltech)

Another notable observation is that the metal content of the Milky Way (which, for astronomers, means all elements heavier than hydrogen and helium) is lower the further out in the galaxy we look. The existence of such heavy elements is, as we know, also a precondition for life, so life conditions in this respect could be less favourable in the outer parts of the galaxy.

Clearly the Sun lies at a suitable location in the Milky Way. Still, this location is the topic of much debate. Since life was created on Earth, the Sun has circled the centre of the galaxy at least fifteen times. As far as we know, the orbit of the Sun around the galaxy is more or less circular, which contributes to stability. But we need to consider that during its long journey, the solar system has come close to other stellar concentrations in a more or less random way. Consequently it has been subjected to the risk of gravitational disturbances or exposed to a supernova explosion. In its path the solar system may also have passed through a nebula, which could dampen the sunlight reaching Earth and also influence the solar system in other ways.

Contrary to general belief, the Sun is actually not a completely ordinary star. Solar-type stars only constitute about 10 % of all stars. Much more commonplace, about 80 % of all stars, are the long-lived red dwarfs. Life conditions are probably worse for planets around these stars, although this is currently a hot topic of discussion. Red dwarfs are smaller than the Sun and therefore the habitable zone available for terrestrial-type planets is considerably closer to the parent star, and also narrower. The proximity to the star means that a planet in its ecosphere is subjected to strong tidal effects which could lead to bound rotation. In many of these cases the planet then has the same hemisphere always pointing towards its star. Normally, this means that one side would be super-heated while the other would stay deep-frozen. This would appear to be an unsuitable environment for life. Currently, theoreticians are investigating whether an energy redistribution and temperature equalisation can be obtained via atmospheric wind patterns.

The Sun varies in its activity, following an eleven year cycle. Sometimes large mass ejections take place which can be directed towards, and collide with, Earth. Normally, the result is only a beautiful display of aurorae, but sometimes the effects can be more serious. For instance, electrical energy systems can be affected and even destroyed. The Earth is largely protected from these effects through its magnetic field, which diverts electrically charged particles towards the magnetic North and South poles. Many red dwarfs also are active, with mass ejections of various kinds, perhaps even stronger than the solar ones. A nearby planet obviously runs the risk of being detrimentally affected.

A Planet With the Right Stuff

When we come to the properties of the actual planet, a set of criteria needs to be satisfied. First and foremost, the planet must lie inside the ecosphere of the star. A too-elliptical orbit would lead to drastic annual climate changes. Earth's orbit around the Sun is only weakly elliptical, with the minimum distance (perihelium) happening on Jan 3, when our planet is at a distance of 147 million kilometres. The maximum distance takes place on July 4 at 152 million kilometres. As a consequence, the solar radiation input varies by about 7 % during the year. However, this is a small effect compared to the seasonal changes (see below).

Naturally, the planet must be of a suitable type; we need a rocky planet with a hard surface, dominated by minerals and metals. In order for such planets to exist, the material from which the planetary system originally formed must have been enriched by heavier elements. As we have seen, the existence of such metals can be verified observationally by spectral analysis of the light coming from the host star.

The rocky planet must also contain a certain minimum mass so that it is capable of sustaining a sufficiently dense atmosphere of some kind. Otherwise, conditions cannot sustain water in liquid form. For example, we have seen that Mars lacks enough gravitation to maintain an atmosphere in the long run.

The Earth: A Double Planet With a Magnetic Field

As a planet the Earth apparently fulfils a number of necessary qualifications. Another, possibly quite unusual, feature of the Earth is the size of our Moon. Several of the planets in the solar system have moons, but typically they are much smaller than their planet. The Moon of Earth is, however, big, about a fourth of the diameter of Earth. Seen from outside it is not unreasonable to consider the Earth-Moon system as a double planet. As such it is the only one in our solar system. How can this have happened? The prevailing theory about the origin of the current Moon tells us that Earth very early in the history of the solar system collided with a protoplanet, about the size of Mars. This cosmic catastrophe subsequently lead to the formation of the current Moon and may also have caused the current inclination of the Earth's axis. A clearly random event, which strongly affected the conditions on Earth in prehistoric times. Double planets may be rare.

Additionally, the inclination of the Earth's axis is stabilised through the influence of the Moon. This inclination is the reason for our seasonal changes, and some believe this may actively have

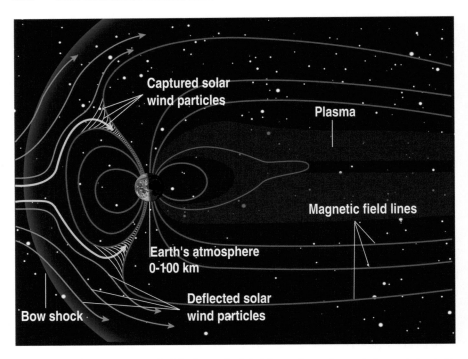

Fig. 10.4 The magnetic field of Earth shields us from harmful particle radiation from the Sun (Peter Linde)

contributed to the advancement of life. The same can be said about the tidal effects on Earth caused by the Moon. It is thus conceivable that our Moon may have had considerable influence on the evolution of life on Earth.

Another important circumstance is that the Earth has a rather strong magnetic field (see Fig. 10.4). Electrical currents in the part of the Earth's interior called the outer core generate this magnetic field. As discussed in Chap. 3, this region consists of floating iron and nickel. Above this layer, magma streams slowly move (see Fig. 3.1). These are also responsible for the plate tectonics that cause the continental drift on the Earth's surface. The Earth's magnetic field may well play a decisive role in the evolution and persistence of life. The field protects us from most of the solar wind to which the Earth is constantly exposed. If the solar wind would arrive unblocked, the ozone layer high in the atmosphere would be in danger of being obliterated. This would allow ultraviolet radiation to penetrate to the ground and destroy the conditions for life. Some scientists even believe that the atmosphere itself would gradually disappear, at least in the case of smaller planets. This fate may have befallen the planet Mars. Today the magnetic field of Mars is nearly gone, but there is evidence that a stronger field existed at an earlier epoch. Of the inner and comparable planets in the solar system, Mercury has a weak magnetic field and Venus an even weaker one. It is not known how typical the strong magnetic field of the Earth really is. Future detailed studies of exoplanets ought to provide an answer.

The Water: A Gift from Heaven?

Oceans dominate the surface of the Earth, spanning about 70 % of the total surface. Their volume is estimated to be more than a billion cubic kilometres. You can rightly ask yourself from where all this water originated, especially since it is considered fundamentally important for the emergence and evolution of life. First of all, we note that water is a common element in the universe. Water consists of hydrogen and oxygen, both of which are ordinary constituents of the universe. Hydrogen dominates all others; 74 % of all mass in the universe is hydrogen. Oxygen is the third most common element (after helium) and is estimated to make up about 1 % of all mass. The big difference is explained by the different origins of the elements. The hydrogen was created at the time of the Big Bang, while oxygen is a product of fusion processes inside heavy stars. It has been added to interstellar space through violent ejections from unstable stars. It is probable that the original nebula from which our solar system was formed contained considerable amounts of water. The question is in which form it existed and where it was located.

It is well-known that water in liquid form on Earth has a boiling point that depends on the surrounding pressure. But how does water behave in space, where vacuum prevails and consequently the pressure is zero? The answer is that in vacuum water can only exist in two forms, either as ice or as vapour.

It is believed that when a planetary system forms, both these forms exist abundantly. The heat from the newly born protostar will vaporize the water within a certain distance. Outside that

zone the water exists as ice. The vapour is gradually swept away
by the radiation of the star, but the ice further away remains, in
the form of comets and a variety of lesser ice-covered bodies. As
we have seen, the critical distance which separates the two forms
of water is called the snow line.

But the Earth is well inside the snow line in our solar system.
The water vapour that originally must have existed in this part of
the newly formed system has probably long since disappeared. You
may wonder where all our existing water actually comes from.
One answer, which today is generally accepted, is that the water
came to Earth (as well as to the other inner planets) quite soon
after their formation as rocky planets. At that time the Earth was
bombarded by lots of ice-bearing bodies. By studying how much of
the hydrogen in the water exists in the form of deuterium (heavy
water) it is possible to establish from where in the solar system
the water originated, since deuterium ratios indicate where the ice
must have originally existed. Part of our information comes from
meteorites landed on Earth, and partly from in-situ measurements
from a few comets. The most recent data came from the spectacular
rendezvous of the comet 67P/Chuyumov-Gerasimenko made by
the Rosetta space probe. These measurements showed a different
hydrogen/deuterium composition from that of Earth. Indications
are that comets may not be responsible for bringing the water to
Earth. Instead ice-covered minor planets are the main suspects.
But we still need to explain why these collisions took place at all,
at such a suitable time in the early history of Earth. One theory
involves the planet Jupiter, which could have played an important
role at this point in time by gravitationally re-routing the orbits
of these smaller bodies. Is this a normal development in a newly
created planetary system, or was it a rare and unique event which
allowed the water to come to Earth?

ETI Exists—But We Never Meet

Let us instead assume that extraterrestrial civilisations actually
do exist. In this case, we need to explain why we have not yet discovered them—or they us. There are a rich variety of hypotheses
to choose from, some more likely than others, most of them quite

imaginative and many entertaining. In his book *If the Universe Is Teeming With Aliens… Where Is Everybody?* (2002), Steven Webb has listed 50 different more or less plausible explanations. We will examine a few of them.

As we now know, the factor L in Drake's equation refers to the lifetime of a civilisation. Unfortunately, we again only have our own civilisation as an example. Several scenarios may lead to a civilisation's destruction, both in terms of human activities and in terms of cosmic disasters. Let us call them inner and outer threats, respectively. We begin by discussing a few self-inflicted risks for annihilation.

Inner Threats

The risk of a nuclear war seems (temporarily?) to have decreased during the last few decades. In the 1960s humanity was not very far from fulfilling the gloomy prediction that all civilisations exterminate themselves sooner or later. Although the situation has improved somewhat, about 10,000 nuclear weapons in at least eight countries still remain, ready to be used. There is little doubt that our civilisation will be erased if these are used on a large scale. In addition to the initial direct devastation, long-term effects such as nuclear winter would be catastrophic. But how catastrophic? In fact, all life would not be completely erased even in the most comprehensive nuclear scenario. Meet *Deinococcus radiodurans*, alias "Conan the Bacterium" (Fig. 10.5). This remarkable extremophile is a primitive organism with the special quality that it is capable of rapidly repairing its own DNA, even after severe radiation damages. Tests show that this kind of bacterium would not be seriously affected by a nuclear war. So the conclusion is that life will not be completely eradicated by radioactivity. The odds for making a comeback in a new and advanced form after a few million years would probably be good. And a hundred million years is a relatively short time from a cosmic perspective.

There are, however, many other self-destruct options. Particle physicists often work with physics at the verge of the unknown. Some worry that the experiments carried out at, for example, the new CERN particle accelerator, LHC, could potentially lead to a

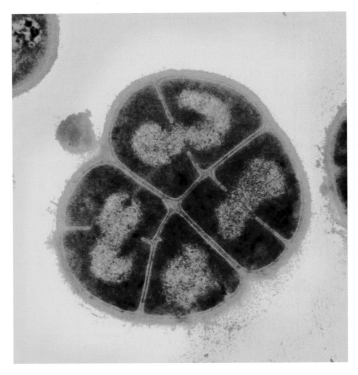

Fig. 10.5 Deinocuccus radiodurans, alias "Conan the Bacterium" (Michael J. Daly/Science Photo Library IBL)

catastrophe. A black hole could be created by mistake. Although originally very small, the black hole might rapidly grow and engulf the entire Earth. There are even some who suggest that such experiments could disturb the condition of what we call vacuum. Modern physics does not view vacuum as an uninteresting emptiness. Instead researchers introduce terms like vacuum-energy and vacuum states. It is conceivable that the current vacuum in space could be unstable and change its state if it is subjected to very high energies. This rift would grow with the speed of light and the whole of the universe might be affected. But by all accounts we can sleep tight even when the particle physicists are running at full throttle. Nature itself has, for billions of years, made similar "particle experiments". The cosmic radiation contains particle energies far exceeding anything generated in human laboratories so far, so such catastrophic effects would have happened long ago.

The Swedish philosopher Nick Bostrom studies so called existential risks. With existential risks he means such risks that can lead to the annihilation of the human civilisation. He is a professor at the Oxford University in England and head of the Future of Humanity Institute. The institute is dedicated to studies of the future development of humanity on very long time scales. These types of studies seem lacking in today's society. Bostrom has written a number of papers on topics like transhumanism, cloning, artificial intelligence, cryotechnology and nanotechnology. He sees enormous potential for technological advances. According to him (and many others), nanotechnology will, in the near future, open up completely new possibilities. Nanotechnology involves building small functional units, machines, literally using only a few molecules. Nanotechnology has potential for a variety of useful things, and promises a brilliant future in important fields such as electronics, medicine and material science. Unfortunately, as is well known, all technologies can also be misused. Derailed nanotechnology is high on Bostrom's list of existential risks. For instance, it is conceivable that hostile governments or terrorist groups could be capable of mass replicating "nanobots" that "consume" the carbon in organic compounds. In a short time, these could invade and endanger the entire biosphere. In extreme cases, this could lead to the destruction of all life in a very short time. Another similar danger would be the application of genetic manipulation, either by mistake or by intention, to create an artificial virus that could rapidly destroy humanity. It is necessary for us to discuss and protect ourselves against these risks, as they are certainly at least on the same level of relevance as the risks involved with utilising nuclear power.

Outer Threats

It is generally acknowledged that the disappearance of the dinosaurs was caused by an enormous meteorite impact about 65 million years ago. In 1994, we got a reminder that such catastrophes may still happen when the comet Shoemaker-Levy 9 crashed into Jupiter. The impact in the Jupiter atmosphere generated an incredible energy output, resulting in marks on Jupiter of a size comparable to the diameter of Earth (Fig. 10.6).

Fig. 10.6 Comet Shoemaker-Levy collides with Jupiter in 1994 (H Hammel/MIT/NASA)

Devastating Earth impacts have also occurred in historic times. Most famous is the Tunguska explosion which took place in Siberia in 1908. It is not definitely established whether its cause was a large meteorite or a fraction of a comet, but the destruction was extensive. The object never reached the ground, but exploded in the atmosphere at a considerable altitude. The size has been estimated at a few tens of metres, but estimates show that it may have developed an energy corresponding to about 1000 Hiroshima bombs. Up to 80 million trees may have been broken at the explosion site, which, however, was at such a remote location that no scientific investigations were carried out there until 1927. It is easy to imagine the catastrophic consequences which would have resulted if the impact had happened in a densely populated area. We were reminded of this fact in 2013, when a large object entered the atmosphere near Chelyabinsk in Russia. Although no direct damage resulted from the impact, the blast wave from the event shattered windows, causing injuries to more than a thousand people.

The threat of asteroid collisions with Earth is real but small. Lesser objects occasionally do come rather close to Earth. In November 2011 the 400 m sized asteroid YU 2005 passed Earth at a distance of about 300,000 km, which is closer than the distance to the Moon.

If an asteroid with a diameter of 2–3 km collided with Earth, it would no doubt have catastrophic consequences for our civilisation. The devastation around the impact point would be enormous. If it fell into the ocean, which is the most probable case, it would generate huge tsunami waves. A collision on land would cause disastrous fires. The fallout from the impact would spread through the Earth's atmosphere, leading to acid rain falling on the vegetation and blocking out sunlight. Experts estimate that impacts of this order may happen with an average interval of about 500,000 years. The earlier mentioned SpaceWatch programme aims at identifying possible threats in advance. However, smaller objects are difficult to detect before they are close to Earth. On May 28, 2012 an asteroid with a size of about 25 m passed the Earth at a distance of only 40,000 km. It had been discovered only five days earlier. However, larger and really serious threats should be possible to identify with sufficient time to allow the option of modifying their orbits enough to miss the Earth and avert the danger.

More potential threats lurk out in space. Certain types of massive stars end their days in gigantic explosions. We call them supernovae, a spectacular example is seen in Fig. 10.7. Such phenomena are though quite rare. The most recent ones observed in the Milky Way were the nova of Tycho Brahe in 1572 and Johannes Kepler's nova in 1604. Such explosions release incredible energies, and the stars increase their brightness by almost a billion times. If a neighbouring star inside a radius of about 1000 light years would explode in this way, it would pose a substantial hazard for biological life on Earth. The gamma radiation could endanger the ozone layer of the atmosphere, after which the ultraviolet radiation from

FIG. 10.7 In 1054 Chinese astronomers observed a supernova. Almost a thousand years later the site of the catastrophe looks like this. We call it the Crab Nebula. Deep down at the centre lies a collapsed remnant in the form of a rapidly rotating neutron star (see also Fig. 11.3) (NASA/ESA/CXC/JPL)

the Sun and energetic cosmic rays unhindered would affect life and life conditions. As a matter of fact, nearby supernova explosions may explain some of the mass extinctions of species that we have observed in the earlier history of the Earth. The risk that our civilisation would become extinct for this reason is difficult to estimate, but is probably quite low.

In fact, supernovae are not the most dramatic event that can happen out in space. So called gamma ray bursts generate even higher energies. Interestingly, they were originally discovered in the 1960s by American satellites monitoring possible nuclear tests in the Soviet Union. A few years ago, astronomers realised that these flashes came from other galaxies at very long distances. So far, none have been detected in our own galaxy—fortunately—because Earth might well be totally sterilised if we got too close to such a phenomenon.

The space dangers do not end there. Instead of Earth being grilled, there may be a risk of deep-freezing. We have earlier mentioned the nebulae and dust clouds from which new stars are born. If, in its motion through interstellar space, our solar system passed through such an area, the solar radiation would most probably be reduced with devastating effects for life on Earth. There are many other causes for dangerous cooling and it is known that Earth has had at least five ice age epochs. There are even indications that the entire planet may have been completely ice-covered during a period about 700 million years ago.

Are We Inside a Wildlife Preserve?

We now return to further external explanations to the Fermi paradox. Were we to ever encounter extraterrestrials, it is generally believed that they would be far ahead of us in their development. This belief is connected to the expected technological life-span for a civilisation (the L parameter). Let us assume that a common value would be as short as 1000 years and that we have advanced 100 years into our life-span, since radio communications became possible. From statistical reasoning, most of the alien civilisations ought to be further into this developmental era, assuming random starting points in time.

It may seem presumptuous to believe that we can imagine the motives they may have for their treatment of a lower-level civilisation like ours. Nevertheless, some try to explain the Fermi paradox with a Zoo—or rather a wildlife preserve—hypothesis. The assumption is that aliens leave us alone much as we do a wildlife preserve; they do not interfere in order to let us develop freely in our own direction. ETI is there, but they never show themselves. Instead, they maintain an observational role. However, the Zoo hypothesis has a couple of rather obvious weaknesses. One is that intelligence on Earth is quite recent, and to observe or maintain a quarantine of single-cell organisms does not seem to be very fruitful. In such a case they must have been in a great hurry: having just detected intelligence they quickly established a natural reserve and made themselves invisible.

Another problem with this kind of scenario is that it cannot be tested. If superior civilisations want to stay hidden, we may never uncover their existence, no matter how hard we try.

They Stay at Home

It is entirely possible for alien civilisations to simply stay at home. Perhaps interstellar travel is judged as too slow and too expensive, even for a more advanced civilisation. It could also be the case that they choose to explore other realities than the surrounding universe. This is not quite as peculiar as it sounds. It is conceivable that an advanced culture may also have advanced capabilities to simulate virtual realities using enormously powerful computers. Such realities may well be more complex and possibly more interesting than the "real" reality. We should keep in mind that our universe is based on a delicate balance between a number of physical constants. Perhaps a supercivilisation can simulate and explore other universes?

Before we dismiss such ideas, it be illuminating to briefly reflect on the development of our own culture. Since radio, television and, in particular, the launch of the internet, many have devoted increasing parts of their time to live pseudo-lives, where the experiences are electronically delivered. Throughout the latest decades, this trend has accelerated in the form of increasingly advanced and sophisticated computer games. Many, not the least young people, spend a large part of their life in front of a computer screen to participate in simulation-based pseudo realities. One example is Second Life, a computerised virtual world, which is estimated to have a couple of million participants.

Social contacts are more and more managed via the internet, making it less and less necessary for real personal meetings. It is not inconceivable that in the near future it will become possible to incorporate into the human brain direct communication with advanced computers and networks. The difference, then, between virtual and real reality will be very small, and many may well prefer to live a rich life in a fantasy world instead of a poor and dull life in reality.

The Technological Singularity

The future prospect of human evolution is a theme that recurs in many science fiction novels. There may be little possibility to predict what humanity will look like in just a hundred years. But it seems likely that we have our destiny in our own hands; the evolution of nature will no more be the decisive factor. Instead, the human capability to modify and change itself is ever increasing. In the more privileged countries, medical advances have already prolonged the average life time considerably. We are on the verge of a new world where humans can be improved both genetically and electronically. Whether we can adapt mentally and socially to these rapid changes is not equally obvious.

Moore's law is well-known in computer science. It describes how many transistors that can be placed on a certain area of a single microchip. The law projects that this number doubles approximately every second year. In practice, this means that the computing, memory and storage capacity increases correspondingly. This law has astonishingly enough been valid over the last forty years, and currently more than two billion transistors can be placed on a single processor chip. Some believe this will lead to the emergence of an electronic super-intelligence in a few decades. Immortality could then be around the corner in a near future. Nick Bostrom discusses a transhumanistic world where mankind is transformed into a completely new physical and intellectual level, using a variety of advanced technologies. Connected to this scenario is the concept of "the technological singularity". This "singularity" implies that intelligence may increase so fast that the consequences cannot be anticipated. An intellectual event horizon may arise beyond which events cannot be predicted or even understood. Perhaps humanity would change its current existence entirely, into something that we cannot perceive at all at this time. Developments like this clearly impact the life span of a "normal technological" civilisation in a totally unpredictable way. We will return to this question in Chap. 13.

Is it possible that ETI did this a long time ago? In this perspective, perhaps the need to colonise the galaxy does not exist or may have become outdated for reasons completely incomprehensible to us.

The Planetarium Hypothesis

A more speculative reason for the termination of mankind, according to Nick Bostrom, could be that our reality is nothing more than a simulation and that someone decides to turn it off. Our reality is assumed to have been created by a super-civilisation. This very same idea has been made popular by the series of Matrix movies. The explanation is sometimes referred to as the planetarium hypothesis. In a planetarium an artificial universe is created in the form of a projection of the sky on the inner walls of a large dome. From the immense development of computing power, with which we are well acquainted from our own world, it is not completely inconceivable that ETI has the power to simulate us and the universe we are living in. But just as we would we able to expose the planetarium sky by a closer look at the projection screen, it ought to be possible to check whether our "reality is real". But that the planetarium hypothesis explains the absence of other extraterrestrials seems a bit—unreal.

Perhaps We Are the First?

In speculations about alien civilisations it is often assumed that they should, after the formation of the stars and galaxies have emerged randomly in time. This is not necessarily true. Astrophysical facts imply that a few billion years must pass before life has any chance at all of being created and evolving.

As we have seen in Chap. 2, organic life is based upon a number of important basic elements: sulphur, phosphor, oxygen, nitrogen, carbon and hydrogen. All these elements (except hydrogen) can only be created in fusion processes inside stars. They then must be thrown away into space in more or less explosive reactions. Much later, they gradually join to form a new nebula, which starts out being enriched in these elements. One realises that the matter of the universe must have been recycled several times in different star generations before this could have happened. When such a nebula starts to fragment to form a star cluster, the preconditions for creating planets suitable for life are already there, with sufficient quantities of the necessary elements. Our Earth is one result. The astronomer Mario Livio points out a few more necessary circumstances, related to the existence of oxygen and ozone in the atmosphere of the Earth. Ozone is a kind of oxygen molecule, with three interconnected oxygen atoms. The ozone appears to be essential, because it protects life from harmful ultraviolet radiation. As we have seen, without this protection the risks for genetic damage are great, especially for land-based life. The initial addition of oxygen to the Earth's original atmosphere was probably created by the ultraviolet radiation from the Sun, which released oxygen from the water vapour that was prevalent at an early stage. On Earth this phase will have lasted for about 2.5 billion years. Simultaneously, as we have seen in Chap. 3, the oxygen content in the atmosphere increased due to aquatic primitive cyanobacteria, which biologically created additional oxygen via photosynthesis. So, with the gradually increasing level of oxygen in the atmosphere, a layer of ozone built up over time. It is estimated that this stage ended about 600 million years ago. After this it became possible for life to enter on to land, and a new and decisive phase for the development of life had begun.

Livio's conclusion is that about 10 billion years are needed before civilisation can actually be created. One may object that the universe according to current estimations is about 13.8 billion years old, which ought to have provided enough time for other civilisations to form before ours, with enough time to colonise the galaxy.

There are more violent explanations to the Fermi paradox. A more brutal solution to the paradox proposes that all civilisations are systematically exterminated in such early phases that they never become competitors to the one existing and ruling super-civilisation in the galaxy. This civilisation could have launched armadas of so called berserkers, highly intelligent machines of the von Neumann type, but with the purpose and capacity to destroy entire planets. It would be unfortunate if, in this science fiction-inspired scenario, they got the idea that Earth is developing into such a potential competitor.

The Great Filter: Is There Hope or Is the Doomsday Near?

The analysis of all the pros and cons in the Drake equation shows that there is an ample supply of good arguments on both sides—regardless of whether you are an optimist or a pessimist. Which scale pan eventually carries the most weight remains to be seen. Unfortunately, the problem is unbalanced in the sense that we can never prove for certain that something does not exist. If, however, a single ETI message shows up on our screens, the matter will be settled.

From the above discussion, it must be admitted that those who say that no other civilisations exist have a lot of indirect evidence to point their fingers at. The pessimist simply states that no trace of another civilisation has ever been found, not on Earth nor in space. They come to the conclusion that we are the only existing civilisation in the galaxy. This means that some factor in the Drake equation must be very close to zero. There could be more than one such factor, but one is enough. This factor sometimes is referred to as the "Great Filter". Nick Bostrom follows Enrico Fermi when he argues that the number of terrestrial-type planets in the galaxy must be enormous. The fact that we do not observe intelligent life elsewhere proves the existence of such a filter. Bostrom considers this as a key issue for humanity. Why is this? Humanity does exist. But the question is whether this filter lies in our past or in our future. Have we passed the crisis or do we have yet to confront it?

Among the plausible filters in our past, Bostrom points in particular to a very low probability for life to be created at all. Another candidate is the advance from primitive life to high-level intelligence, something which may also be very improbable. Most of us are looking forward to the prospect of scientists soon being able to do on-site searches for life somewhere in our neighbourhood, and hopefully then find traces of life. Not so Bostrom. He earnestly hopes that we will not find any traces of life on Mars or any other place in the solar system. If this happens, he considers it as proof that the filter—and thus the crisis—looms in our future. Some kind of major event or process is waiting to exterminate us. Moreover, this will happen in a rather near future, before we ourselves start to colonise the galaxy. On the other hand, he argues, if the filter existed in our past history, our future could be very bright indeed, with almost unlimited possibilities for evolution at all levels.

The Doomsday Argument—The End of Humanity or Paradoxical Intellectual Exercise?

As we have already discussed, the factor L in the Drake equation, the expected life-span of a civilisation, is the most decisive one and also the most unknown. But perhaps there is a way to predict how long our civilisation will exist. In 1993 Richard Gott, American astrophysicist, published a hypothesis that came to be known as the Doomsday argument. He says that on purely logical and statistical grounds it is possible to predict when the human civilisation will cease! His remarkable line of reasoning is certainly well worth taking a closer look at.

The starting point is a generalisation of the Copernican principle, which we have touched upon in the previous chapter. Its conventional interpretation is that we humans and our place in the universe is in no way unique. Gott's generalisation is an extension of this principle, linked to time. Our location in time is equally non-specific. When we make an observation here and now, it gives us a possibility to arrive at certain conclusions, provided that the observation can be considered as being done randomly in time.

Gott illustrates his arguments with a personal illustration from real life. In 1969 he saw the Berlin Wall for the first time. This event could be considered as a random event in the time history of the Berlin Wall. In this case, you can argue that there

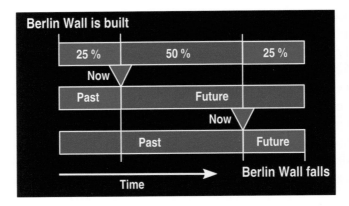

Fig. 10.8 Gott's visit to the Berlin Wall predicted its life expectancy (Peter Linde)

is a 50 % probability that his visit took place during the middle half of the wall's lifetime (see Fig. 10.8). If he visited in the beginning of this period, evidently 3/4 of its lifetime would remain. If he saw it at the end of the middle half, 1/4 of the lifetime would remain. The wall was erected in 1961, eight years before Gott's visit. After some basic reasoning and calculation, it turns out that this means there is a 50/50 chance that the remaining lifetime would be between 2 2/3 and 24 years.

Via a purely mathematical recalculation this can be expressed as a 95 % probability that the remaining lifetime was between just over a week and 32.5 years. As we all know, the wall was demolished in 1989, twenty years after Gott's visit. So this type of logic at least seems to have worked for the case of the Berlin Wall. Could it be applied to mankind, and extended to predict its fate?

Let us begin by performing a simple (and classical) thought experiment. We assume we have in front of us two large urns containing marbles. One of them contains ten numbered marbles, the other contains a million numbered marbles. It is not known which is which. Now we randomly take one marble from one of the urns. It turns out that it has the number seven. From which urn is it more probable that this marble has been taken? Of course from the one with ten marbles, because the probability in favour of this is 99.999 %.

Now we instead imagine that the urns represent two hypothetical human races, all members also numbered, in this case in order of birth. We belong to one of the species but we do not know which. Our number in this case turns out to be 60 billion. We realise that this is a reasonable number if this urn (=race) contains, for instance, 100 billion individuals. To belong to the other urn would be highly improbable if it contained, say, 10 trillion people.

Following Gott's argument, if we randomly note that our number is 60 billion, we must assume that it is considerably more probable that the human race (including future members) contains 100 billion individuals than if it would contained 10 trillion. It is believed that so far about 70 billion people have existed. From this you can postulate that the human existence is drawing to its end. Nick Bostrom, who analyses Gott's arguments in detail, uses statistical methods to calculate that this corresponds to a probability of 95 % that between 1 and 2700 billion individuals will live in the future. In order to translate a numbering of people into a time period, you need to estimate the future development of the Earth's population. A plausible (and necessary) development is that the population in the near future will stabilise at about 12 billion. In this case, according to the calculations of Bostrom, humanity will, with a probability of 75 %, disappear within 1400 years.

Now we certainly wonder to what extent these mind-boggling and frightening results are generally accepted. Sure enough, there is a lot of criticism against this kind of logic. One weakness is the actual definition of what is meant by a random time-independent observation. It clearly needs to be done by a sentient being. And when did man become sentient? May we include humanity's possible successors in the form of super-intelligent machines? How much are Gott's ideas actually connected to reality? There is clearly room here for everyone's thoughts. And the philosophers are already battling it out.

What Is the Conclusion?

We find that several of the explanations for not yet having observed any ETI have to do with our assumptions of how the ETI thinks. To attempt to extrapolate from Earth-based motives and morals is certainly to tread on thin ice. A further difficulty with this type of explanation is that there must be a strong consensus and unity among the motives of the aliens over very long time. It is enough that a single ETI decides to colonise the galaxy in order to offset these explanations.

Other hypotheses suffer from the weakness that they cannot be tested, even in principle. And all hypotheses suffer from chronic lack of information. But at the same time it is also obvious that some of this lack of information will decrease in the near future. The continued research on exoplanets will give some answers.

So, how do the pessimists use the Drake equation? Our optimistic evaluation from the previous chapter was

$$N = R. \cdot F_p \cdot N_e \cdot F_l \cdot f_i \cdot f_c \cdot L => 7 \cdot 0.5 \cdot 3 \cdot 0.5 \cdot 0.1 \cdot 0.5 \cdot 50,000 = 13,125$$

The pessimists might instead perhaps assign the value of 0.01 to f_l (creation of life), 0.01 to f_i (emergence of intelligence) and 5000 to L (life-span of a civilisation). Then the result instead becomes

$$N = 7 \cdot 0.5 \cdot 3 \cdot 0.01 \cdot 0.01 \cdot 5000 \approx 1$$

This result would fit rather well into what we actually know to be true, namely that there is one reasonably advanced civilisation in the Milky Way.

While there are many reasons to explain why we have not yet seen traces of ETI, there is only a single explanation to why we would: namely, that what happened to initiate intelligent life on Earth also happened elsewhere in our galaxy, perhaps many times. If intelligent species are commonplace but leave no traces, will humanity, too, meet a similar destiny? Or if an alien community is finally revealed, will humanity adapt to it?

11. SETI: The Search for Extraterrestrial Intelligence

In the beginning of 1930s the American radio engineer Karl Jansky was given the task of investigating if there were any sources of interference that could impede short-wave radio communication. For this mission he designed a special radio antenna which, supported by Model T-Ford wheels, could be rotated in every direction. Aside from the usual disturbances from thunderstorms, he also noticed a strong noise coming from a certain position in the sky. In 1933 he published his findings in a classic report in which he demonstrated that these radio waves actually emanated from an area far away outside the Earth. In fact, they came from our own galaxy, the Milky Way. This was the birth of radio astronomy, and Jansky might well have received a Nobel prize for his discovery, but he died rather young, just 45 years old.

At this point, when technology allowed transmitting signals over large distances and receiving radio radiation from space, it was not a far-fetched idea that artificial signals were already out there, ready to be picked up.

The concept of SETI appeared in the beginning of the 1960s. The acronym stands for Search for Extraterrestrial Intelligence. Early on, the term was often spelled with a "C", which stood for Communication (with...). However, many considered this an overoptimistic attitude, so the "C" soon became a more moderate "S". Astronomers realised that communication with an alien civilisation, if it was possible at all, would lie very far ahead in the future. Hypothetical response times would be counted in many years, considering the enormous distances involved.

The question of whether to broadcast or to listen was, however, easy to answer. As we saw in Chap. 9, the life span of a civilisation becomes very relevant for this discussion. Our civilisation has matured sufficiently for interstellar communication only in the last 100 years. Even if we assign a rather pessimistic 1000 years

© Springer International Publishing Switzerland 2016
P. Linde, *The Hunt for Alien Life*, Astronomers' Universe,
DOI 10.1007/978-3-319-24118-0_11

as a typical technological life span, on the average about 90 % of the civilisations would be more advanced than ours. This assumes that civilisations emerge randomly during long periods of time. Since it is considerably easier to listen than to broadcast, and with more immediate results, it was very reasonable to replace CETI with SETI. The more realistic goal now is to detect ETI-signals, i.e. extraterrestrial intelligent signals.

How Difficult Is SETI?

The difficulties of catching intelligent signals from space are huge. First of all, you must assume that we are not alone in the universe, you have to be an optimist. After that, you need to imagine how an alien civilisation is thinking, one that may well be thousands of years ahead of us in development. Clearly, we are already hopelessly lost in what can only be considered as speculation. In any case, we face a very difficult technological challenge, and we must take advantage of the most advanced technology available in order to have any chance of success. Finally, we need to arm ourselves with supreme patience. However, there is a great risk that funding agencies for SETI projects lose interest when presenting a scenario of this kind. Understandably, donors prefer results in the shorter run. As we will see, this has already happened several times. Governmental interest in SETI has varied, but it is not an exaggeration to say that interest has generally been quite weak. Therefore, budgets for SETI projects have normally been quite thin. Overall, they have often survived thanks to contributions from private individuals and friendly companies. The SETI research is pursued by certain kinds of scientists who stake their careers on something with no guarantee for a positive result.

So what is the problem? Let us begin by trying to guess what we can expect out there.

Kardashev's Super Civilisations

It has long been speculated upon how the signature of an alien civilisation could be recognised. It is natural to primarily concentrate on civilisations that are reminiscent of our own, although possessing a much higher technology level. Such a civilisation does not necessarily want to communicate with lower level beings (like us), but it may have left traces from higher technological activities. The correlation between technological level and increasing energy consumption is often taken for granted, even though this may contradict our current ideas of a sustainable society. Based on this reasoning, the Russian astronomer Nikolai Kardashev defined a kind of scale that was intended to describe different levels of civilisations, as measured by their energy consumption. In the 1960s, he divided extraterrestrial civilisations into three main levels:

I. Civilisations with capabilities of channelling and exploiting all the energy resources of their own planet (for example, a civilisation capable of exploiting the entire energy input reaching the planet from the host star). In the case of Earth, this total infalling power is on the order of 10^{17} W, which corresponds to almost a billion nuclear power plants. Our civilisation is evidently not yet in this category.

II. Civilisations with the capability of channelling and exploiting the complete energy output of its host star. An example of this would be to enclose, at a suitable distance, the entire star in kind of shell, perhaps by redistributing the mass of its planetary system. The astronomer Freeman Dyson discussed this idea as early as 1960, and today such hypothetical systems go under the name of Dyson spheres. The Sun, a rather typical star, has a total luminosity corresponding to 4×10^{26} W. This is four billion times more than level 1.

III. Civilisations with the capacity to channel and exploit the complete energy output of their galaxy. A rather typical galaxy, like the Milky Way, contains about 300 billion stars. How such energy levels could in practice be utilised is difficult to even speculate about. Nonetheless, as we will soon see, there are scientists actively working to try to identify this type of super civilisations.

Information Beacons or Leakage?

What kind of extraterrestrial signals can we hope to detect on Earth? There are two different scenarios. One envisages civilisations that have the intention to contribute to—and accelerate the development of—newcomers. The second involves attempting to eavesdrop on the normal communications inside or between superior civilisations.

In the first case we could probably assume that they would like to make it easy for us. They ought to have previously passed through the technological level that we currently possess. Perhaps they consider themselves similar to parents who want to educate their children. Therefore, they adapt themselves to our technological standards. They may have, for instance, deployed specially designed information beacons with the sole purpose of disseminating valuable information. The question then becomes in what form would they choose to distribute that information. According to the physics that we know, it seems obvious that electromagnetic waves in some form would be the natural choice. One important reason is that they propagate with the speed of light. Plausible alternatives could, for example, be gravitational waves or neutrino pulses, but in addition to being much more difficult to produce, they would also be extremely more difficult to detect. Information transfer in the form of a kind of particle radiation seems unlikely, as many of the particles like protons and electrons (which both exist in the cosmic radiation) have an electrical charge. This means that they follow curved paths, affected by the galactic magnetic field. Additionally, particles are easily obstructed by the interstellar medium. So, for reasons to which we will return, radio signalling is, in many ways, the most obvious choice.

Information beacons should have the suitable and positive property that they maintain their transmissions during long periods. A message should run for a long time and be frequently repeated. Once you locate a signal of this kind, you should be able to rely on finding it again, with the possibility of studying it in detail. We can also somewhat optimistically assume that such messages have been transmitted a long time ago and have had enough time to reach the Earth through the cosmic depths. We cannot predict to which of the Kardashev levels such a civilisation belongs. It may be worthwhile to also search distant galaxies. Perhaps one single type III civilisation is enough for us to see a signal? We simply have to ignore the fact that it might be millions or billions of years old when it arrives.

The other option for finding civilisations is through their "normal" information leakage. This is considerably more difficult.

We hear a lot of talk about Earth having exposed itself via the radio waves that leak out into space from our radio and television broadcasts. And to some extent this is correct; our transmissions have undoubtedly reached out in spherical globe with an approximate diameter of 100 light years. But if we look closer to this you see that purely physical laws imposes strong limitations. Not even a super civilisation can deal with the fact that at, for instance, at a distance of 10 light years, there would only be isolated photons to catch from the leakage. It would demand radio telescopes with diameters of many kilometres to be able to form any kind of signal. It may not be an impossible undertaking for an advanced civilisation, but definitely very resource demanding.

Another aspect to consider is that radio leakage may well be a temporary stage in the development of a civilisation, thus it is limited not only in space but also in time. From our own experience during the latest decades, development in this direction has already begun. More and more of our communications are now more effectively transmitted in cables, fibre-optic, or others. The radio leakage into space is getting smaller and smaller.

What really makes a difference is whether a signal is directional or not. The television noise from Earth can be described as an isotropic transmission, which means that on the average it is transmitted in all directions simultaneously. The intensity of such a signal will be reduced inversely proportionally to the square of the distance. A simple calculation shows that a transmission that can easily be monitored at the edge of our solar system will be about 250 million times weaker when it has reached a distance of 10 light years. Directed transmissions are quite different. A directed signal focuses the transmission into a beam which is only gradually weakened by distance. One example is radar signals, which are transmitted in order to reflect back an echo, which then can be detected. For this reason it is assumed that powerful military radar systems probably broadcast transmissions with the highest probability of detection at interstellar distances. If an alien civilisation uses a similar technique, with great luck we may be able to pick it up. But most probably it would not be repeated, and thus it would always be very difficult to confirm an observation of this kind.

The search for ETI signals is clearly much more difficult than finding a needle in a haystack. We need to use hyper-sensitive equipment to look in exactly the correct direction at the correct time and on the correct channel. Nevertheless, an ongoing listening takes place, which we are now going to take a closer look at.

How Do You Find An Artificial Signal?

The search for signals from extraterrestrial civilisations picked up momentum when radio technology developed at the beginning of the twentieth century. A couple of the greatest contemporary engineers, Nikola Tesla and Guglielmo Marconi, both declared enthusiastically that they had received signals from Mars. However, the sensation quickly died out. The signals could probably be explained by atmospheric phenomena or possibly by natural radio emissions from the planet Jupiter. The understanding of radio waves coming from the universe was, at this stage, rudimentary and it was not until considerably later that scientists began to understand their causes. As we have seen, the first real radio telescopes were built in the 1930s and led to the discovery of radio emissions from hydrogen gas in the Milky Way. The second world war gave a strong push for further radio technology advancements, which later could be put to good and more peaceful use in astronomy. In 1964 Arno Penzias and Robert Wilson made their famous (and Nobel prize award-winning) discovery of the cosmic background radiation. It became a cornerstone in the emergence of modern cosmology, which is based on the Big Bang theory.

Radio emissions constantly reach us from outer space. Simultaneously, we also use them for communication on Earth. It is quite possible that other civilisations are doing the same. But radio waves are electromagnetic radiation which exists in many forms, each characterised by the wavelength. The shortest (as well as most energetic) wave types of radiation we call gamma, X-ray and ultraviolet radiation, respectively. These are followed by visible light, infrared radiation and finally radio radiation. Out of these, only visible light and radio waves reach Earth's surface (see Fig. 5.12 in Chap. 5); the rest is blocked by the atmosphere and has to be observed using satellites.

Which channel (i.e. frequency band) should we use in order to maximise the probability of receiving a cosmic message? As we have seen, the electromagnetic spectrum is extremely wide.

There are billions of channels to choose from, and we would appreciate limiting the frequency range that we need to monitor. We should keep this fact in mind if we are discouraged by the quite poor search results so far. In order to limit the number of channels, there are a couple of useful facts to take advantage of. Signals with frequencies lower than 1 GHz (wavelengths longer than 0.3 m) drown in the galactic background noise. On the other hand, frequencies higher than about 30 GHz (wavelengths shorter than 1 cm) do not penetrate the atmosphere of the Earth. Nonetheless, enormous numbers of possible channels remain. Can we restrict this even further? That became one of the first problems that the SETI project tried to solve.

The First Pioneers

Many astronomers (and politicians!) have often viewed SETI with scepticism, and even today this is still true to a certain extent. For many years, searching for "little green men" was considered a frivolous way for scientists to spend their time. But serious SETI research actually began in 1959. That year, a short but epoch-making paper was published in the esteemed journal *Nature*. It was written by physicists Philip Morrison and Giuseppe Cocconi. In it the authors described the feasibility of using microwaves for interstellar communication. They made the bold claim that we are surrounded by civilisations that have been broadcasting messages for a long time. They asserted that these messages were just waiting to be picked up by worlds like ours, worlds capable of listening for signals. They also proposed that such civilisations would make it as easy as possible for us to receive and interpret a message. For instance, they would realise that our Sun provides a suitable environment for a life-bearing planet.

Cocconi and Morrison suggested that the frequency 1420 MHz or somewhere near it ought to be a natural choice for interstellar communication. This is a very well-known frequency, since it is broadcasted by hydrogen atoms, the most common element in the universe. At least it would be a natural choice in the directions pointing away from the galactic plane where the background noise would be smaller. They pointed out that an artificial signal should show a Doppler effect due to the orbital motion around the host star of the originating planet. They also identified a few nearby stars suitable for monitoring. The paper ended with the often quoted remark: "The probability of success is difficult to estimate; but if we never search the chance of success is zero". This would become something of a motto for the emerging SETI research. The paper gave a respectability to this new area. Gradually, SETI was established as its own scientific field. It thus became possible, at least for the more daring scientists, to actually set to work searching for other civilisations. Soon after, it was realised that another frequency, 1660 MHz from the OH molecule, could also be interesting. Since hydrogen (H) and OH together form water (H_2O), a fundamental condition for life as we know it, the frequency range between 1429 and 1660 MHz became known as the "water hole". What could be more natural and fitting for different civilisations than to meet at the water hole?

Soviet scientists were also active within the field, especially in the 1960s and 1970s. Nikolai Kardashev and Iosif Shklovsky were among the big names. Shklovsky was not afraid of controversial hypotheses and suggested, among other things, that one of the Martian moons, Phobos, could be a hollow satellite and thus of artificial origin. As the reason, he cited the strange orbital motion of the moon. This theory was eventually dismissed. In spite of the cold war, Russians and Americans kept in touch, bridging the political gulf, by organising some joint conferences.

With time SETI became more and more an American phenomenon. High-tech equipment was necessary, and in this field the Soviet Union gradually lagged behind. However, finding and convincing financing authorities was always the main problem. But in the USA a growing community of astronomers began to take the SETI discussion seriously. Many of these were soon to be counted as the real pioneers of SETI. In addition to the previously mentioned Cocconi and Morrison, there were new names like Frank Drake and Carl Sagan.

Project OZMA

Frank Drake was the first to attempt a systematic search for messages from other civilisations. In 1960 he started Project OZMA with the intention of studying some of the nearest solar-type stars.

In 1959 Drake had come to Green Bank in West Virginia, USA, where they were in the process of building large radio telescopes. The site had been declared as a radio quiet zone in order to provide the best possible preconditions. The intention was to build the foremost radio observatory in the world. One radio telescope with a diameter of 26 m had just been finished (Fig. 11.1). Drake realised that the telescope had the potential to catch signals from a hypothetical civilisation within a radius of about 10 light years. He managed to get support for the idea to set up a monitoring project and so the development of suitable receiving equipment began. A fortunate circumstance for Drake was that his telescope could be equipped with a completely new type of amplifier, at least ten

FIG. 11.1 The radio telescope at Green Bank which Frank Drake used for the first attempts at listening for intelligent signals from cosmos (NRAO/AUI)

times more sensitive than previously available. Meanwhile, quite independent from Drake's work, Cocconi and Morrison published the previously discussed paper. It worked as a further encouragement. Additionally, the astronomer Otto Struve had recently become head of the observatory. Struve was a very conservative individual of the old school, raised during the Russian czarist regime. But he was also one of the few well-known authorities at this time who publicly declared his conviction that extraterrestrial civilisations did not only exist but were also numerous. So it was with great expectations that the first search for alien intelligent signals was started in April 1960. Drake tells us:

"On the first day of Project Ozma, I set the alarm clock for three, got up groggily, and went out into the fog and cold which was to be my regular morning greeting for about two months. At the 85-ft telescope, the operator would turn the telescope so that I could climb into the metal can, not much bigger than a garbage can, which was at the telescope's focus. There I would sit for about 45 min twiddling the micrometer adjustments on the parametric amplifier, talking to the telescope operator, so that it was doing the right thing."

The first target was the solar-type star τ Ceti at a distance of twelve light years. During the entire morning the group was staring at the chart recorder which registered the measurements, but only typical noise was recorded. After that the telescope was pointed towards another solar-type star, ε Eridani, at a distance of ten light years. After five minutes something happened, the chart recorder made gigantic deflections, eight times a second. Could it be that easy? Or did the signal have a terrestrial origin? Checks were made, the telescope was moved off the star, the signal disappeared as expected, the telescope was moved back—and the signal did not reappear. The excitement was great and the speculations were many. The observations of ε Eridani were continued uninterrupted but nothing more happened. The cause of the signal was unknown for ten days. Suddenly the signal reappeared. This time a secondary backup antenna was pointed to another part of the sky, unfortunately the signal was heard also there. It evidently had a terrestrial origin, in all probability the signals came from a high-flying aeroplane passing by. The project continued the studies for another two months using a large number of radio channels. No suspicious artificial signal was ever detected. But at least the SETI research had taken it first wobbling steps.

The Dolphins Meet at Green Bank

The pioneering observations of Frank Drake and the audacious paper by Cocconi and Morrison became the starting point of a new era. In November 1960 a small and select group of scientists and other experts gathered for a conference at Green Bank. The meeting kept a very low profile; no announcement had been made and no publication followed the meeting. The topic of the conference, opportunities for establishing contacts with other worlds, was still dubious enough to endanger the scientific reputation of the participants. Only eleven persons were present and all them would, with time, become famous in SETI research. They covered different competencies. One of the participants, the biochemist Melvin Calvin, was told during the meeting that he had won the Nobel prize in chemistry. A rising prominent figure within SETI, Carl Sagan, also participated. John Lilly, psychoanalyst and philosopher, had just published a work on the intelligence and communicative capacity of the dolphins. For this reason the little group named themselves "The Dolphin order".

The conference became a classic. For the first time, a number of experts made positive statements about the possibilities for communication with alien civilisations. In addition, for the first time attendees articulated the SETI issue in a mathematical formula, the so called Drake's equation, which we have discussed in detail in Chap. 9. The meeting closed by declaring that attempts to communicate with highly developed civilisations in the Milky Way should be undertaken. Estimations for their number ranged from less than a thousand to more than a billion. A call was made to instigate a decisive and rigorous search using very large radio telescopes, very powerful computers and a 30 year long patience.

CTA-102: False Alarm

In the Soviet Union technology and science was a priority, and during the cold war there was an incessant race with the West in these respects. During the 1960s, the Soviets invested large resources in radio astronomy. As we have seen, Nikolai Kardashev was active in SETI matters and in 1964 he published a controversial paper in a Soviet professional journal, where he claimed to have received signals from alien super civilisations. In particular, signals had been received from two objects designated CTA-21 and CTA-102. These radio sources were already known but now Kardashev had caught signals that he claimed were clearly periodic—a sure sign of artificial origin (see Fig. 11.2). Estimations had shown that the distances to the objects were rather short, a hundred light years, and Kardashev suggested that the signals must have come from very advanced civilisations. Follow-up observations were carried out which corroborated the regularity of the signals. The Soviet news agency TASS even released a telegram about the super civilisations which became a sensation all over the world.

However, this information was received with scepticism from the scientific community. The argument that the signal was periodic was not convincing. In Fig. 11.2 the inlaid dashed curve is very approximate, data do not necessarily have to be interpreted as periodic. Additionally, American astronomers disputed the distance determination to the object, since a new type of mysterious radio source had been discovered a few years before. After thorough searches, visual equivalents to these sources were identified, i.e. objects that were located in the same places as the radio sources.

This was also successful in the case of CTA-102. By using the world's largest optical telescope at the time, the 5 m telescope

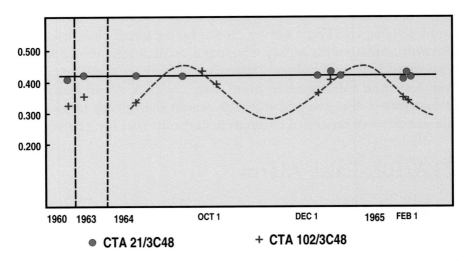

FIG. 11.2 Observations by Kardashev of the radio sources CTA-102 and CTA 21. CTA-102 shows variations which may be interpreted as regular (Peter Linde)

at Mount Palomar, it finally became possible to obtain a spectral analysis of the light from the object. It turned out to have an enormous red shift (see Chap. 5), which corresponded to a receding velocity of about 31,000 km/s. In accordance with the theory of the expanding universe this immediately placed CTA-102 at an enormous distance—about a billion light years away. Detailed studies also revealed that the signals from CTA-102 were hardly periodic. The mysterious radio signals were shown to come from a new type of objects that would become known as quasars. The explanation as to how it is at all possible to see these objects from such distances proved to be yet another cornerstone of fundamental astronomical discoveries in the twentieth century. Quasars are embryos of galaxies which existed billions of years ago, with incredible energy emissions driven by black holes. But that is another story.

The Woman Who Did Not Win the Nobel Prize

During the autumn 1967 the young student Jocelyn Bell was working assiduously on her thesis in astronomy. She had gotten the opportunity to use a recently inaugurated radio telescope at the Mullard radio astronomical observatory near Cambridge in England. The telescope had been in operation since the summer and was basically designed to investigate scintillation in the interplanetary medium. This is less remarkable than it sounds. The radio waves from radio sources in space pass through our solar system before they are registered in the radio telescopes on Earth. In the solar system, charged particles emanate from the Sun (the so called solar wind), and these affect the radio waves in a way corresponding to what happens to the starlight as it enters the Earth's atmosphere. The stars seem to twinkle, and in the same way the radio sources twinkle in radio telescopes.

For this reason the radio telescope which Jocelyn Bell used was equipped with an advanced receiver with high time resolution, in order to be able to follow rapid changes in the radio signal. In the large material that Bell was collecting, there were some signals that seemed to repeat themselves with great accuracy.

They appeared to be located in the same position in the sky, which ruled out various Earth-related explanations like radar signals and satellite signals. The phenomenon merited a closer study, and Bell started a new series of observations in November. At the end of the month things were settled. There was clearly a radio signal that varied very regularly, much like a time signal. Every 1.3373th second a radio pulse arrived. Such an extreme regularity was definitely not expected from a natural phenomenon. Partly as a joke, the object was originally dubbed LGM-1, where LGM meant "Little Green Men". It caused great excitement: could this be the first message from an alien civilisation? But Jocelyn Bell was irritated: "Here was I trying to get a Ph.D. out of a new technique, and some silly lot of little green men had to choose my aerial and my frequency to communicate with us".

The investigations continued. Soon another pulsating radio source was discovered, this time in quite another constellation and with a different period between pulses. A pattern started to emerge, and the theory that the signal came from alien civilisations began to fade away. The signal had no Doppler shift, a change in wavelength you would expect from a transmitter on a planet orbiting a star. It had to be something else. What had been discovered did, in fact, have a natural explanation, in many ways was as sensational as the explanation involving the little green men. In the beginning of 1968 the journal Nature published an epoch-making paper. Anthony Hewish, Jocelyn's supervisor, was the first author name, Bell the second. The newly discovered objects were named pulsars, from the words pulsating star, and turned out (yet again) to be the gateway to studies of a completely new and exotic type of stars, that we today call neutron stars (see Fig. 11.3 and Box 11.1).

Jocelyn Bell quickly became a celebrity when media got hold of the news. At a dinner speech given many years later she sarcastically remarked: "The excitement was great after the publication of our discovery. I had my photograph taken standing on a bank, sitting on a bank, standing on a bank examining bogus records, etc. Meanwhile the journalists were asking relevant questions like was I taller than or not quite as tall as Princess Margaret and how many boyfriends did I have at a time?"

Fɪɢ. 11.3 The pulsar in the Crab Nebula is a neutron star which rotates 30 times a second. The image is a combination of an optical image and an X-ray image (NASA/CXC/HST/ASU/J. Hester et al)

The discovery of the pulsars was epoch-making, and in 1974, for the first time, the Nobel prize was awarded to astronomers. Anthony Hewish was awarded half the price for the discovery of the pulsars and Martin Ryle the other half for his revolutionary new designs of radio telescopes. But the one who actually made the discovery, Jocelyn Bell, was left out. Many think that the Nobel committee made a big mistake. However, Jocelyn herself did not agree, when she modestly commented: "I believe it would demean Nobel Prizes if they were awarded to research students, except in very exceptional cases, and I do not believe this is one of them. Finally, I am not myself upset about it—after all, I am in good company, am I not!" Incidentally, the pulsar discovery was only mentioned in an appendix of her doctoral thesis.

Box 11.1: Pulsars and Neutron Stars

The existence of neutron stars had been postulated by theoreticians since the 1930s, and in 1967 the theories were confirmed by actual observations. A neutron star is a kind of collapsed star, one out of three possible end stages for a star which has used up its nuclear fuel. The other two are white dwarfs and black holes. When the nuclear fuel has been depleted, there is no longer any force directed outwards, keeping the object stabilised. Gravitation dominates completely, and the star drastically shrinks in size. For a neutron star, this results in a collapsed object with extreme density. The collapsed star becomes 100,000 times smaller than in its original state, perhaps with a diameter no larger than 20 km! The only remaining force that can withstand the enormous pressure is the nuclear force itself, which rules inside the atomic nuclei.

During the collapse, the neutron star also begins to rapidly spin up. For a normal star, like the Sun, the rotation period could be about 25 days. Neutron stars have been shown to rotate 300 times a second! Figure skaters make use of the same phenomenon, as they increase their rotation by withdrawing arms and legs as close to the body as possible. Physicists refer to this as maintaining angular momentum.

During the collapse also the magnetic field of the star becomes concentrated to enormous levels. The result is not unlike a kind of dynamo, and the rapidly spinning object in the dense magnetic field emits strong radio pulses. Similar to a light-house, observers detect one pulse per rotation. This phenomenon was what Jocelyn Bell had discovered.

Project Cyclops: Futuristic in the 1970s

Nevertheless, the search for intelligent signals carried on in novel ways. Although SETI research at the beginning of the 1960s was generally accepted and had gripped the interest of the public, it took another ten years before anyone was ready to make a new solid attempt in the USA. During this time the discussion in scientific circles was intensive and many new ideas saw the light of the day. Gerritt Verschurr became the man who started new observations in 1971. He had a 100 m radio telescope at his disposal and a much more sensitive receiver than Frank Drake had used eleven years earlier. Verschuur monitored nine stars during two years, but only with a total effective observing time of thirteen hours. He found nothing.

In the 1970s a new development in SETI research took place when NASA decided to join the search. With the powerful economic muscles of NASA, this led to new resources and a potential for more advanced studies. During 1971, NASA sponsored a study leading to a concept called project Cyclops. A group of 25 experts in different fields worked on the study for half a year. A report was published, filled with technical calculations and advanced ideas. The plan aimed for a very ambitious goal. The tone was set at its beginning, where Frank Drake was cited:

"At this very minute, with almost absolute certainty, radio waves sent forth by other intelligent civilisations are falling on the Earth. A telescope can be built that, pointed in the right place, and tuned to the right frequency, could discover these waves. Someday, from somewhere out among the stars, will come the answers to many of the oldest, most important, and most exciting questions mankind has asked."

The group proposed building an enormous radio telescope. A thousand interoperating individual radio telescopes, each with a diameter of 100 m, were to be located inside a radius of 5 km (Fig. 11.4). The sensitivity would be so high that it was expected to be capable of eavesdropping on "regular" radio transmissions from other nearby civilisations. The cost for a fully completed system was estimated at about 10 billion dollars, comparable to the Apollo project, which at this time was busy putting men on the Moon.

Because it was sponsored by NASA and contained detailed technical descriptions, the Project Cyclops report carried a lot of weight, becoming a milestone in SETI research. Antennae were designed on paper, as well as receivers and transmitters, and a thorough investigation was made into the problem of analysing the signals. A number of conclusions and recommendations were listed. They were unanimous that SETI was well worth doing and that the pay-off, although uncertain and with results far into the

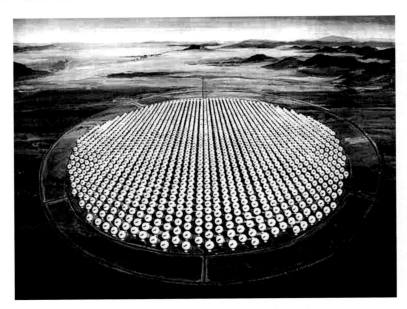

FIG. 11.4 Project Cyclops was intended to contain about one thousand radio telescopes working together, each with a diameter of 100 m. This is an artist's conception of how they could be arranged in a circle with a diameter of approximately 10 km (NASA)

future, would be well worth the investment. The "water hole" was recommended as the frequency region of choice.

Cyclops was never built. The size of the project was far beyond what NASA could sanction. But they could still support less ambitious projects, and this would also be of great importance. In 1979 NASA decided to support a considerably down-scaled plan. Two search strategies would be implemented at the same time. Firstly, a large number of individual stars would be monitored. Secondly, a large survey of the entire sky would be attempted. The whole project became an official NASA project under the collective name Microwave Observatory Program (MOP).

MOP ran into problems from the very start. By 1979 the project was awarded "The Golden Fleece" by American senator William Proxmire. This award was annually announced by him, and was given to governmental projects that, in his opinion, unduly and meaninglessly wasted tax payers money. In 1982 he succeeded in having the US congress cut off the funding for the project. Scientists protested this decision. Carl Sagan made a personal visit to Proxmire and put together a petition signed by many

FIG. 11.5 Big Ear, the radio telescope that caught the signal. Today this area has been transformed into a golf course (NAAPO)

of the world's leading scientists, among them seven Nobel laureates. That helped. The congress continued to support the NASA SETI programme for another decade.

Meanwhile SETI continued to establish itself as a separate field of research. An important step was taken in 1982 when a new commission was formed within the International Astronomical Union (IAU) with the designation "Bioastronomy".

The 1977 Wow Signal

At 04:16:01 in the morning of August 16, 1977 a strong radio signal reached the Earth from the direction of the constellation Sagittarius. It was registered by the large radio telescope Big Ear in Ohio, USA (see Fig. 11.5). The observatory had been looking for interesting radio signals during the course of four years. The search had become routine, and was managed by volunteers. When the signal came, the telescope was unmanned and all observations were done and documented automatically. A few days later,

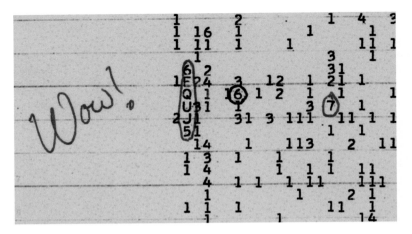

FIG. 11.6 The Wow-signal from 1977. On the computer printout the powerful signal is circled with red ink (NAAPO)

astronomer Jerry Ehman as usual was sifting through the latest results, given in the form of crude computer printouts. He did not expect anything out of the ordinary. To his immense astonishment he found evidence of an enormous signal, the likes of which he had never seen. He marked the signal in the printout with red ink and underlined his surprise by adding the word "Wow!" in the margin (see Fig. 11.6). This word later became the designation of this incident—perhaps the most exciting one which has ever been registered in the search for other civilisations.

At this time computers were still primitive and documentation and storage options limited. Ehman and colleagues had programmed a kind of simplified diagram where the power of the radio signal was assigned a number according to the principle that the higher the signal, the higher the number. Just in case the number scale would not suffice, the computer was programmed to also use letters. This was fortunate since the wow-signal was much stronger than anyone had foreseen. The symbols 6EQUJ5 represented a very strong and clear signal. It caused a huge sensation. The signal was analysed through and through, but the highest priority was to observe it again. For more than a month Ehman repeated attempts to catch the signal again, but to no avail.

So what was it that made this signal so interesting? It had several of the characteristics you would expect from a signal broadcasted from an extraterrestrial civilisation. First of all, it was received at a frequency (1420 MHz) suitable for interstellar communication and forbidden for use by terrestrial transmissions. In addition, the signal was narrow in frequency, only 10 Hz, which means that it did not contain many other frequencies at the same time. This is similar to the way ordinary radio broadcasts work: you tune in to a certain frequency on your radio in order to listen to a single station. In nature the opposite is more common, radio sources transmit at several frequencies at the same time, subject to the laws of physics. The appearance of the signal strongly indicated a space origin, not a nearby terrestrial transmitter. Is also appeared distinctly limited in time. The design of the radio telescope was such that every object would be observed twice, with a time interval of three minutes. This, however, did not happen; the signal was only observed once.

Many speculations about the origin of the mysterious signal have been put forward. Perhaps most interesting is to study Jerry Ehman's own ideas, documented twenty years after the event. He notes that planet or satellite explanations could be ruled out. Aircraft move across the sky and this would have affected the shape of the signal. A considered possibility was that a ground-based radio transmission could have spilled out into the frequency in question and then been reflected from some satellite or junk in orbit around the Earth. However, closer investigation showed this to be very unlikely. In fact, no credible man-made explanation could be found, although military signals of some kind could not be ruled out completely. It was also noted that the signal came from a position in the sky that lies close to the plane of the Milky Way, an area with high stellar density.

Did the signal contain any information? Unfortunately this question remains unanswered since Big Ear was not equipped to study rapid time variations in the signal. But several other projects have since focused on studying the position in the sky from where the signal originated, among them several more powerful and modern than Big Ear. Unfortunately, the signal has never reappeared.

Big Ear met with a grim fate. The university which owned the ground sold it in 1998, and the telescope was demolished in order to facilitate an extension of a nearby golf course. But fortunately the search for intelligent signals continued in other ways.

Millions and Billions of Channels

In the 1980s the technology on which SETI was so dependent accelerated further. Although it was possible, by logical reasoning, to limit the frequency range, the remaining number of channels suitable for communication still seemed close to infinity. The improving technology not only concentrated on increased sensitivity but also on listening in on many channels in parallel. In 1981, Harvard physicist Paul Horowitz developed a portable spectrum analyser. His device could connect easily to a radio telescope, and used sophisticated hardware and software to analyse the data being received. SETI requires heavy computing capacity, since the analysis uses advanced mathematical computations (known as Fast Fourier Transforms, FFT) to identify interesting signals. The Horowitz spectrum analyser was known as Suitcase SETI, and could simultaneously analyse 131,000 narrow-band channels. In 1983 observations with the 26 m Harvard/Smithsonian radio telescope in Massachusetts, USA, began. The project was called Sentinel and was active until 1985. It was then replaced by project META, an acronym meaning Megachannel Extra-Terrestrial Assay. Now the number of simultaneously searched channels had reached 8.4 million. Horowitz still headed the project, which was partly funded by film producer Steven Spielberg. After five years of study, Horowitz and Sagan reported 37 candidate events, none of which however were repeated. A similar instrument, META II, was set up near Buenos Aires, Argentine, in order to scan the Southern part of the sky. This instrument has been upgraded a couple of times and is still running.

But millions of channels were not enough. The successor was called BETA, the "B" now meaning "billion". This project started in 1995. But four years later it abruptly ended, when the 26 m radio telescope was damaged in a storm to an extent that made it irreparable. The media hinted that ETI had struck in order to prevent a possible discovery…

NASA Goes On-line—And Off-line Again

For historians, October 12 is a special date. On this day in 1492, Christopher Columbus first set foot in the new world. Although a coincidence, it seemed symbolic that this date was also the day when mankind started serious searches towards new horizons. Exactly 500 years after Columbus, NASA initiated two studies in parallel. In the Mojave desert in California, a 34 m telescope began scanning the entire accessible sky. At this time, 16 million channels could be investigated concurrently, compared to the 50 channels Big Ear could manage fifteen years earlier. The goal was to catch stronger signals, perhaps especially sent from another civilisation with the intention of facilitating easy reception. This surveying search also included any conceivable broadcast from hypothetical super civilisations in distant galaxies.

At the same time, scientists at the gigantic Arecibo telescope in Puerto Rico instigated a new, directed search. With its 300 m diameter, Arecibo stood as the world's largest radio telescope (see Chap. 8). Almost 1000 stars were selected to be investigated with the highest available sensitivity in the frequency range 1–3 MHz, including the "water hole".

Less than a year after starting these ambitious projects, a shock descended on all SETI researchers and sympathizers. The US congress suddenly and unexpectedly cut off all funding for the project. This time, the Nevada senator Richard Bryan led the crusade against the NASA SETI project, stating:

"The Great Martian Chase may finally come to an end. As of today millions have been spent and we have yet to bag a single little green fellow. Not a single Martian has said take me to your leader, and not a single flying saucer has applied for FAA approval."

After 23 years of work and 60 million dollars invested, Congress killed the project less than a year after it started. This time, not even Carl Sagan and thousands of other scientists were able to save it. But although the NASA effort was incomplete and short-lived, it still had great importance for continued SETI research. In the future activities SETI became much more professional, and experts put to good use the most advanced equipment available. In addition, parts of the collapsed NASA project could be salvaged and reused.

Project Phoenix

Thus, in 1993 NASA abandoned SETI research, creating kind of a void. However, some of this void was filled through other efforts. By 1984 the SETI Institute had been founded with the intention of supporting projects and conducting research connected to SETI. The institute is located in Mountain View, California. One of the founders is Jill Tarter who until recently was the research leader there. Tarter was the model for the movie character Ellie Arroway in the movie "Contact", based on Carl Sagan's science fiction novel of the same name.

The SETI Institute succeeded in pursuing NASAs cancelled program of monitoring a number of stars in detail. It was now appropriately given a new name, Project Phoenix, rising as it were from the ashes. Much of the NASA equipment could be taken over, and in 1995 observations resumed. A list of almost 1000 stars was constructed. Most of them were solar-type stars, within a distance of 200 light years and older than 3 billion years. The receiver and analysis system for Project Phoenix was mobile and could subsequently be used on different telescopes. The first one utilised was the Australian Parkes telescope (see Fig. 11.7), with a diameter of 64 m. Eventually, access was also provided to the previously mentioned Arecibo telescope, where short observing times were allocated during a number of years.

In 2004 the project was completed. At that point 800 stars had been monitored in short periods but in more than a billion channels. The project leader, Peter Bacchus, commented that "we live in a quiet neighbourhood". He was however not pessimistic, but rather looked forward to new observations.

The SERENDIP Project

The other part of the old NASA programme, a systematic survey of the whole sky, also had a successor, christened SERENDIP. The word itself alludes to making an unexpected discovery, but it is also an acronym with the long interpretation "Search for Extraterrestrial Radio Emissions from Nearby Developed Intelligent Populations". A second group of SETI enthusiasts had been formed in 1980, with Carl Sagan as one of the founders. It was called the Planetary Society and became, together with the SETI institute, a general driving force. The society advocated the search principle to look over the sky without focusing on any particular target. When the NASA effort was cancelled, the society took it upon themselves to continue the surveying.

FIG. 11.7 The Parkes radio telescope in New South Wales in Australia. It has a diameter of 64 m (Stephen West Diceman)

SERENDIP had already begun in 1979, when a first version of their spectral analyser was built. True to its acronym, SERENDIP took advantage of a different mode of operation. It participated as a secondary instrument in ordinary astronomical observation runs, thus improving chances of getting observing time. The SERENDIP team did not control the way the telescope was pointing, which is not a serious drawback for a project which aims to survey the whole sky.

After the initial spectrum analyser, SERENDIP has advanced with several new generations of hardware and software. With each successive step, its technical performance has improved incredibly, due to the digital revolution which continued to follow Moore's law (see also Chap. 10). The commercial development within digital processing has consistently led to higher capacity at a lower price. For SERENDIP this meant that the number of channels which simultaneously could be searched increased from 100 to 2 billion! Since 1990 this system has been used at the Arecibo telescope in order to obtain maximum possible sensitivity. Searches have been implemented for signals both around 420 MHz (70 cm) and around the previously discussed "water hole".

Incidentally, SERENDIP's version IV got a second life at a large radio telescope outside Bologna in Italy (SETI-Italia). But at Arecibo the fifth version of the system is currently operating. Now a multi-receiver is used,, enabling observations at seven sky positions simultaneously. This means that browsing the sky goes faster. In addition, collaboration with other astronomers with similar interests also improves efficiency. Many interesting signals have indeed been detected, but so far none that can be clearly attributed to another advanced civilisation.

SETI@Home: A Revolution and a Success

The SERENDIP project has been running since 1979 in various incarnations—and is still running. Its spectrum analyser performs gigantic amounts of computations. These are done live, in real time, and any results are immediately available, allowing for a quick follow-up if needed. In spite of the huge computing capacity of the system, it is still only enough for a relatively superficial analysis of the data. The same data that SERENDIP uses could also be studied in more detail to extract even more fine-grained information. But a deeper analysis would only be possible if access to even larger amounts of computing power were made available.

In 1994 David Gedye, a project leader at the company Starwave Corporation, realised that it might be possible to get assistance from the general public. At this point in time, personal computers were becoming commonplace in homes, and due to the simultaneous rapid development of the emerging internet, many of them were available for communication. The home computers often sat idle, and Gedye's idea was to gain access to the unused computer power of thousands of small computers. Over a couple of years, Gedye worked out the idea in detail and developed the necessary software. Instead of the trivial screen saving programmes which the PCs were running most of the time, useful analysis of radio astronomical data could be performed. In 1999 they were ready to launch a project that would soon prove to be revolutionary, due to its enthusiastic reception from the general public. It was given the name SETI@Home.

How Does SETI@home Work?

Each day, the Arecibo telescope receives large amounts of data for SETI analysis. By 1999 data levels had reached about 35 GB per day. Since Arecibo is rather isolated, located in the jungle of Puerto Rico, it possessed no fast connection to the internet, so in the beginning it was necessary to ship magnetic tapes with stored data by mail to the SETI@home centre at the Berkeley University in California. There, the data collection was divided up into many smaller pieces that could be transferred via internet to volunteers all over the world.

The software that was developed contained a small but highly specialised analysis program that could easily be installed in ordinary PCs. The whole process was automatic. The computer of the user only needed to be connected to internet during limited periods. In the first step, the PC received the data to be analysed. Subsequently, the program executed only at times when the computer would otherwise be idle. Usually, the program in total needed a couple of hours to process one data segment. After completing the analysis, the results were automatically sent back to the SETI@home centre in the form of a short report, which was correlated to other collected results. Then the user received another data segment and the procedure was repeated. In Fig. 11.8 we see an example of the screen saver in action. For more details about the computations, see Box 11.2.

SETI@home was an immediate success. The project management had hoped for perhaps 100,000 participants, but after only three months more than a million individuals had offered the use of their computers. Everyone hoped to make the revolutionary discovery, a clear and distinct signal, far above the noise level.

Initiated and driven on a shoe-string budget, the project's intention was to run for a couple of years. But in 2005 the project shifted into a new phase. Funding was obtained to develop a new

FIG. 11.8 An example of how the SETI@home screen saver may look like. Data for 107 seconds of observations are successively analysed. Usually the analysis only reveals noise on the screen (SETI@home)

and more general platform for distributed projects. The result was the BOINC system. With this new software platform it became possible for other scientific groups to set up similar computing projects. Today there are in excess of 50 different projects to choose from, in addition to astronomy and physics within fields like medicine, biology, artificial intelligence, environment and mathematics.

This naturally led to increased competition for SETI@home and the number of participants has decreased. But the total computing capacity has nonetheless increased due to the advances in computer hardware. Currently the project counts about 1.5 million users.

Clearly, as a new concept SETI@home became a huge success. The basic idea of involving the general public in scientific projects in this concrete way has proved very important. Today the total number of computers which have contributed is approaching ten million. And many interesting and suspicious signals have indeed been registered. Some of these have been followed up by special observations, but so far an artificial origin has never been verified. In February 2003 an unusual radio source was discovered which was given the prosaic name SHGbo2 + 14a. This object does not fit any known natural phenomenon, but on the other hand it does not show obvious signs of being artificial. So we are still waiting for that perfect SETI signal.

Box 11.2: Which Computations Are Made By SETI@home?

The computations made in the user's computer are quite extensive. The data received for analysis consists of digital signals from a 2.5 MHz wide frequency range, centred on the special channel of hydrogen, 1420 MHz. Each part contains a 107 second observation in a 10 kHz subchannel. The program searches for different types of signals. Primarily, the assumption is made that the signal would be very narrow, essentially containing a single frequency. This is the most efficient way to broadcast great distances with a signal. In addition, there are very few natural radio sources with this property. Another important aspect is the timing properties of the signal. It should first increase in intensity during six seconds after which it should decrease during the same interval. This is explained by the example of the Arecibo radio telescope, which is fixed in place so that each object needs twelve seconds to pass through the focus of the telescope (known as "beam" in radio jargon). Electrical interference of terrestrial origin does not share that property.

Another characteristic under study is whether a signal varies periodically. An artificial signal containing information should display such variations. Finally, the software must also take into account the Doppler frequency shift, expected from the source's presumed orbit around a planet or a star.

SETI software analyses data using "Fast Fourier Transforms" (FFT), a mathematical operation that allows singling out individually strong frequencies from an otherwise weak digital signal. In total, this computer-intensive analysis makes about 3 trillion computations for each received data segment. Each signal found above a certain threshold is automatically returned to the SETI@home centre.

SETI in Visual Light

Many scientists doubt that SETI is all about radio signals. Granted, we know very little about how alien civilisations think, or the possible motives behind any kind of message. When the SETI research was initiated, radio technology was already well established while the laser technology was in its infancy. It was originally assumed that any form of artificial light signals would completely drown in interfering light from the parent star. Therefore it took a while before anyone seriously considered searching for optical signals. But when the radio searches did not supply any immediate results, a few researchers started to look at this possibility in some detail.

At this point, the potential of using laser light for communication became gradually clearer. A rough calculation showed that combining a high-power laser with a 10 m optical telescope yielded an infrared signal a thousand times more powerful than the Sun's. This type of signal could be observed at distances of hundreds of light years. We can do this now, and thus it is not a very bold guess that a more advanced civilisation could do the same and much more. In addition, at the receiving end, advanced optical sensors became available, suitable and adaptable for optical SETI observations.

SETI with laser light has many advantages. The frequency of the laser light is about a million times higher than for radio waves. Therefore, much more information can be sent in a shorter time interval, for example using extremely intensive but very short pulses. A laser beam can be focused much better than a radio transmission and is less prone to interference than radio signals. Clearly, the use of laser light is very efficient.

But a well-focused laser beam also has disadvantages from a SETI point of view. In order to register anything at all the receiver must be located somewhere in the very narrow cone where the signal is propagated. Such a message would probably have to be directed to, and intended for, Earth. Another disadvantage is that visible light is attenuated when it passes through gas and dust in the interstellar media, and this poses a problem for long range messaging.

Optical SETI makes use of various search strategies. The American exoplanet researcher Geoff Marcy has been looking for very narrow so called emission lines (see Fig. 5.7) in the spectra of stars. The idea is to find such optical signals—using very high spectral dispersion—so narrow that they could only be generated using laser technology. This kind of observation assumes that the laser transmission is more or less continuous, i.e. that the transmitter is operating all the time. In 2002 Marcy reported that he had investigated 577 nearby stars without any positive results.

Another strategy utilises the properties of lasers to create intensive but extremely short light pulses. A message of this kind would perhaps only last a few nanoseconds, but during that time the laser would be much brighter than any star. Fascinatingly, the technology to detect such signals exists today. Hypersensitive detectors are routinely used, capable of detecting, counting and locating single photons. Even though they operate in the optical range, these instruments do not create images of anything, and therefore telescopes with high optical precision are not needed. Instead the term "light buckets" is used. This type of highly time-resolved observation also lends itself well to parallel observations from different sites. A laser pulse hitting Earth should be observable more or less simultaneously all over Earth, as long as the source is visible.

Such investigations have been carried out in small scale at several US universities and observatories during recent years (e.g. Harvard, Berkeley and Lick). A few thousand neighbouring stars have been searched for short-lived information flashes, but so far nothing has been found that does not have a more mundane explanation than extraterrestrial signals.

Going back to the Kardashev's civilisation levels, there could also be reasons to look for level III civilisations. As we recall, such civilisations would utilise the entire energy output of their galaxy. Although it is far from simple to understand how this can be achieved, it could possibly affect the galaxy in an observable way. Most probably, a single type III civilisation in a galaxy would be sufficient. For instance, if a majority of the stars in a galaxy were surrounded by Dyson spheres, intercepting the radiation of the stars, would this affect the appearance of the whole galaxy, seen from cosmic distances? Some scientists believe it would. Dyson spheres cannot be completely dark; they would re-radiate heat at longer wavelengths, typically in the infrared.

Groups in Sweden and the USA are currently testing this hypothesis. From a sample of 2500 galaxies, the Swedish group finds infrared excess in just a few candidates, less than 0.5 % of their sample. Their criterion is based on a 75 % energy reprocessing. Using data from the infrared survey satellite WISE, Griffith et al recently published a more extensive investigation, looking at approximately 100,000 galaxies. They find only 50 candidates where 50 % of the total stellar output of the galaxy output may be explained by energy reprocessing by a type III super civilisation. None of these studies are of course conclusive; remaining candidates may well have astrophysical explanations. But indications are that type III civilisations are either extremely rare or that there is something wrong with the basic assumptions of the Kardashev scale.

Project Argus: Opportunity for Amateurs

Many people today engage themselves actively in the SETI research, not just professionals with big telescopes with sophisticated electronics backing them up. The SETI League is an organisation founded in 1993, just after NASA withdrew its SETI support. The SETI League primarily initiates and supports projects that aim to engage radio amateurs all over the world, improving coordination to secure an uninterrupted monitoring of the whole sky. Ordinary people with an interest in technology can participate with their own parabolic aerials 3–5 m in size. With the advanced but cheap technology of today, it becomes realistic to obtain a sensitivity equal to the Big Ear telescope that made the classical Wow observation in 1977. The project goes under the name Argus. Its goal is to have 5000 of these small radio telescopes dispersed over the world, making coordinated observations. That goal has so far not been achieved. A breakthrough to the general public, as dramatic as the one made by the SETI@home project, has not yet happened. It should be kept in mind, though, that it is one thing to use a few key and mouse clicks to set up an ETI search programme on your home computer, and quite another to finance and fund a radio telescope, which inevitably needs quite a lot of technical knowhow. So the 5000 participant goal has proved to be unrealistic. But the project is still alive with almost 150 operating radio telescopes in 27 countries on all continents.

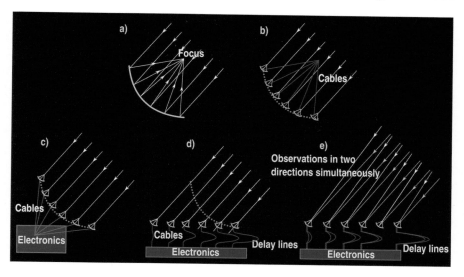

FIG. 11.9 A large radio telescope can be replaced with many smaller working together in a phased array (Peter Linde)

The Allen Telescope

The dream of a giant Cyclops telescope has remained a dream—so far. But the idea of combining a smaller number of telescopes, collaborating to become one large telescope, has certainly survived and prospered. This is the case both in optical and radio astronomy. As mentioned in Chap. 7, at least three major projects are now building extremely large optical telescopes using the technology of segmented individual mirrors. Radio astronomy has long utilised the same principles which, incidentally, are simpler to apply at radio wavelengths. Costs are decreasing since it is possible to mass produce many smaller parabolas. Through advanced technology and data processing it is now possible to attain the same or even better specifications than with a corresponding single dish. As the engineers tend to say, it is cheaper to replace steel by silicon. The technical term is phased array.

Figure 11.9 schematically shows how this works. At the top (a) we see the principle for a single large dish, like the gigantic Arecibo telescope. The radio waves are reflected by the antenna to a single point, the focus, where the receiver is located. Now imagine that the large parabola is replaced with a number of smaller dishes with shorter foci (b). If you then send the collected signal by cables to the same focal point as before, everything should be equivalent. The point is that all cables must have exactly the same length, the travelling distance for the signals must be equal. But if you send the signals in cables they might just as well go directly into a control room (c). And when you think about it, the smaller parabolas do not necessarily have to located along the original curvature, you can just as well put them all on the ground and instead compensate for the resulting difference in travelling distance by using cables of different lengths (d). You can even change the position of the sky you are watching by simply manipulating the length of the cables (e). The final beauty of this technique is that the cables themselves can be replaced by electronic delay lines under computer control. This provides incredible versatility, as we will soon see.

The plans for a very large telescope consisting of several small parts had long been discussed at the SETI Institute. The idea originally came from veteran Frank Drake. Drake aimed to cover an area of 100×100 m, so the telescope initially was called "The One Hectare Telescope" (1HT). The possibility of dedicating such a large telescope to SETI research alone was enticing. In 2001, Paul Allen, co-founder of Microsoft, donated funds so that the design work could be intensified. Three years later construction began. Now the plan included 350 smaller dishes, each with a diameter of 6 m. Instead of building one single specially designed giant telescope, commercial companies were used to mass produce the small units at a considerably lower cost. The donation was enough to fund the first stage, consisting of 42 telescopes, and in 2007 these were inaugurated (Fig. 11.10). The telescope was renamed the Allen Telescope Array (ATA).

Distributed radio telescope offer further advantages. Such a system is easily extendible, it is possible even for private individuals and companies to finance new units at $150,000 each and then give their name to the telescope unit. Funding to build all the planned 350 telescopes is currently not available, but is hoped to arrive in the future.

FIG. 11.10 Close-up view of one of the antennae of the Allen telescope, which consists of a number of interconnected disks, each with a size of 6 m (Colby Gutierrez-Kraybill)

The SETI Institute now has access to a specially designed and dedicated instrument for SETI research. For financial reasons, they must share observing time with other astronomy research projects. This works very well since the telescope can listen to all stars in its large field of view (about 2.5°) at the same time as other astronomers who are, for example, studying a galaxy. It can also be used for other purposes, for example surveillance of satellites and miscellaneous space junk.

The Allen telescope currently monitors more than 2000 planet candidates that the Kepler project has identified so far.

They follow the axiom that if you know where a planetary system is located, chances to find something should be greater. The liability of weak funding was emphasised for ATA in 2011 when, for economic reasons, the telescope had to be mothballed for a time. However, it was soon online again.

Current SETI Research

What is the current status of the SETI research? More than 50 years of searching has so far not resulted in any confirmed contact. So is SETI simply a wasted effort on an impossible project? Of course there are many, like the infamous senator Richard Bryan, who believe this. Jill Tarter, one of the foremost key figures of SETI, instead has this answer:

"We do not go to bed disappointed, we rise in the morning, eager to continue. Our 50 years of searching is equivalent to scooping a single glass of water from the Earth's oceans to examine it for fish. It is an experiment that could work—but if it fails, the correct conclusion is that there was inadequate sampling, not that the oceans are devoid of fish. Today, our searches are getting exponentially better. If we are looking for the right thing, it will take only a few decades to conduct a search that is comprehensive enough to be successful or to yield conclusive negative results. In this field the number two is the crucial number. We count one–two–infinity. We are all looking for number two."

At the same time SETI research continues to be rather controversial. Across the world, very little governmental funding is allotted to SETI and, compared to other fields, very few scientists are actively involved in it. It is primarily private financing that supports ongoing projects. This may not necessarily be evil, but rather a trend in our time. Another example along the same lines may be future manned spaceflight, which will probably be developed and dominated by private and commercial motives. But SETI support currently depends on cultural and philanthropic values, not on commercial ones.

Within SETI there are a few large private organisations that have had a substantial influence on its development. The SETI Institute, The Planetary Society and SETILeague are the three largest, all based in the USA. These organisations raise funding that can be allocated to sponsor various initiatives. Jill Tarter believes that this is the way forward: she asserts that the general public must be involved to an even higher level than it currently is, both in the form of direct participation in the projects and in the form of economic support.

In spite of its negative results, SETI research has progressed both in width and depth. During the past 50 years, our technology has evolved with breathtaking speed. The digital computer revolution and developments in micro-electronics play a major role. The sensitivity of the investigations has increased enormously. It is now feasible to make new kinds of observations that were completely impossible in the 1960s, especially concerning optical SETI. And a completely new factor has come into play. We now know with certainty that there are lots of planets around other stars. This fact has invigorated SETI and the enthusiasm for it; there are now clear targets to aim at.

The Breakthrough Initiative

In July 2015 a major new initiative to rejuvenate and modernise SETI research was announced. Russian billionaire Yuri Milner will during ten years spend $100 million to support the monitoring of a million of the stars closest to Earth, as well as the hundred nearest galaxies. The entire plane of the Milky Way will also be examined, as well as its central parts. In these fields there are billions of stars. Two of the world's largest radio telescopes will be used, the 64 m Parkes telescope in Australia and the 100 m telescope at Green Bank, West Virginia, USA.

The project, which is launched under the name "Breakthrough Listening", is backed by many well-known scientists and leading personalities. Examples are physicist Stephen Hawking and Astronomer Royal Martin Rees as well as SETI researchers Frank Drake, Geoff Marcy, Seth Shostak, Jill Tarter, Nikolai Kardashev and Paul Horowitz.

The new initiative will take advantage of state-of-the art technologies, both in hardware and in software, allowing for a much wider and deeper search than previously possible. About

10 billion frequency channels will be simultaneously monitored, resulting in a daily data reception rate corresponding to a full year of previous SETI efforts. This huge amount of data will also be made available to the general public for further analysis via software building on the SETI@home ideas. Also included in the project is support for optical SETI, i.e. search for optical laser-based signals.

A separate project, "Breakthrough Message", is aimed at finding the optimal way of sending a message from Earth. This project has the form of an international competition, with the goal of formulating a message which can represent humanity and the planet Earth. The intention is not to actually broadcast a message, something which is considered controversial by many. "Breakthrough Message" focuses on analysing the best way to formulate a message and to illuminate the ethical and philosophical aspects of communicating with an alien civilisation beyond Earth.

New Domains

Other domains of signals are now accessible for investigation, mainly in the time-resolved domain. We discussed the short optical pulses above. Similar investigations are also conceivable in the radio domain. A recent project goes under the name AstroPulse. Instead of assuming that that ETI signals must be narrow bandwidth they instead propose that ETI messages can be sent in the form of wideband pulses. If so, completely new analysis algorithms are needed. Basically, AstroPulse uses the same data that is being analysed in SETI@home. Unfortunately a new problem materialises if you are searching for short-term and broadband pulses. Even if the pulse has originally been transmitted as a single giant pulse it will, after passing through the interstellar medium, have been spread out time-wise. You may believe that all radio pulses are propagated at the speed of light but this is only true for the vacuum. In a medium (even though extremely thin) signals move somewhat more slowly, though higher frequencies are faster than lower ones. Incidentally, the same is true for light. A radio pulse that has passed many light years on its way towards Earth will for this reason be spread out in time. AstroPulse aims at trying to reconstruct and discover such signals. The computer power needed for this is enormous, but it becomes manageable by using a modernised version of the SETI@home concept and distributing among hundreds of thousands of users.

Interestingly enough, visual observations made by humans have once again become relevant and important. The idea is to take advantage of human brain's capacity to distinguish weak patterns in complicated images. Even advanced computer algorithms have difficulties in discovering the unexpected; here the human brain still excels. Under the collective concept Zooniverse, the general public is invited to participate in a number of scientific projects. It can be considered as the counterpart to the previously mentioned BOINC project. In BOINC the computers of the general public are used, while in Zooniverse their brains are used.

One of these projects is called SETILive. You participate by studying incoming information from the ATA telescope. Most of the analysis is done by advanced computers, but the fraction of the observations that contain interference of terrestrial origin is left for human visual inspection. The observer applies certain criteria to classify the signals in the images. In addition to the attraction of the (microscopic) chance of finding a genuine ETI signal, you in this way also help to train the automatic algorithms in order to facilitate automatic identification of interference. Since 2011, SETILive can also be run via an app both on Android and iPhone smartphones. Now anyone can help with SETI anytime and anywhere.

New Radio Telescopes

The dream of super large radio telescope still lives. ATA is aiming at 350 units but is entirely dependent on private funding. However, other giant projects are in the making. Although they are not intended for SETI, they may nonetheless play a very important role. Perhaps the first SETI discovery will not come through the expected channels at all, but more as a chance discovery while

FIG. 11.11 The world's largest large radio observatory, ALMA, is now becoming operational. It is located at an altitude of 5000 m, near Chajnantor in the Atacama desert, Northern Chile (ESO/NAOJ/NRAO)

observing other targets. Similar things have happened many times in the history of science.

The most impressive effort within radio astronomy so far is the ALMA telescope (The Atacama Large Millimeter/Submillimeter Array). ALMA contains 66 telescopes with 12 and 7 m dishes. It is located at an altitude of 5000 m in the Atacama desert in Northern Chile. The dry air there allows radio observations almost all the way out to the infrared part of the spectrum (Fig. 11.11).

A very different giant radio telescope is LOFAR (Low Frequency Array) which recently began operations (Fig. 11.12). Here scientists detect low frequency radio transmissions using very simple types of aerials, more similar to TV antennae than radio dishes. About 20,000 such antennae are situated in different places in Europe. With LOFAR the phased array technique has more or less been perfected. Although none of the antennae are movable, it is possible to decide in which direction an observation

FIG. **11.12** One of LOFAR's 96 dipole antennae, at the Onsala Space Observatory outside Gothenburg, Sweden (Peter Linde)

will be done simply by means of extremely advanced signal processing.

The real super telescope, with the name SKA (Square Kilometer Array), is planned to be ready around 2025. When realised, it will far surpass the old Cyclops dream. About 3000 units of 15 m dishes with a total area of a square kilometre will work together, probably dispersed on two continents, Africa and Australia. The experiences from ATA, LOFAR and ALMA will set the scene for this radio telescope, which will have amazing and superior capacity.

The Historic Day

What happens if and when we verify that the SIGNAL has actually arrived? Naturally, this has been repeatedly discussed, and there are concrete proposals for how such a fantastic and pervasive discovery would be handled. However, formal and binding international agreement does not seem to exist. The closest we come to this is United Nations Office for Outer Space Affairs, UNOOSA, hardly one of the more well-known organisations of the UN. At least the matter has been mentioned, mainly through the proposals and suggestions from other organisations. Primarily, it is the International Academy of Astronautics (IAA) that has formulated a concrete plan of action. A dedicated group of experts, the SETI Permanent Committee, takes charge of this. It contains almost 50 members from 13 different countries, half of them Americans. Among them are several of the most well-known scientists from the SETI Institute, including Seth Shostak, who is the chairman of the committee. In Europe, Italy is well represented, probably motivated by their ongoing SERENDIP experiment, which is running at the Medicina radio observatory close to Bologna.

A set of principles has been worked out and were adopted in 2010 at an IAA meeting in Prague. It contains eight points and can be summarised thus:

- SETI experiments will be conducted transparently, and its practitioners will be free to present reports on activities and results in public and professional fora.
- In the event of a suspected detection of extraterrestrial intelligence, the discoverer will make all efforts to verify the detection, using the resources available to the discoverer and with the collaboration of other investigators, Such efforts will include, but not be limited to, observations at more than one facility and/or by more than one organisation. There should be no premature disclosures pending verification.
- If the verification process confirms that a signal or other evidence is due to extraterrestrial intelligence, the discoverer shall report this conclusion in a full and completely open manner to the public, the scientific community, and the Secretary General of the United Nations. A formal report will also be made to the International Astronomical Union (IAU).
- All data necessary for the confirmation of the detection should be made available to the international scientific community.
- If the signal is electromagnetic, observers should seek international agreement to protect the appropriate frequencies via the International Telecommunication Union (ITU).

- The IAA SETI Permanent Study Group will assist in matters that may arise in the event of a confirmed signal, and to support the scientific and public analysis by offering guidance, interpretation, and discussion of the wider implications of the detection.
- In the case of the confirmed detection of a signal, no response will be sent without first seeking guidance and consent of a broadly representative international body, such as the United Nations.

As we can see, the recommendations include a very open approach in accordance to good democratic principles. One might question whether these would be followed in practice. One of the more problematic issues is the one concerning actions for a possible answer. Political and military interests may well play a dominating role, as may religious ones. Ray Norris, Australian radio astronomer and member of the SETI committee, warns that religious extremists may want to send their own message to the newly discovered civilisation. Despite the fact that any possible repercussion would come many years in the future, it would certainly be a worrying development.

12. Journey to the Stars

Will mankind ever be able to visit the stars and other planetary systems? The dream has been alive since early man looked at a sparkling star-filled sky for the first time, many thousands of years ago. As we have seen, the advances in natural sciences have led to a gradually but greatly improved understanding of the structure of the universe. There is good reason to believe that in a few decades we will be capable of colonising planets and moons in our own solar system in earnest, as well as taking advantage of opportunities like mining the asteroids. We do not see any direct show-stoppers for this progress, at least not in terms of the technology required. Any problem would be more likely of a political or economical nature. But commercial issues will soon play a decisive role. In the solar system we can find raw materials needed for the future industrial development.

However, to try to reach for the stars is an entirely different thing. The sad and undeniable truth is that distances to these celestial bodies are truly enormous. In Chap. 1 we took a look at the distance scale of the universe. With light-time as our measure, our solar system is about 10 light hours in diameter. The distance to the nearest star, Proxima Centauri, is 4.2 light years. This star, in the constellation of Centaurus, is the faintest in a triplet of stars that have the common name α Centauri. A simple comparison between interplanetary and interstellar distances shows that it is about 260,000 times farther to the nearest star than to our Sun! To bridge such a distance with manned or unmanned spacecraft currently seems like an insurmountable task. Nonetheless, during the last decades studies have been made to more closely examine the challenge and to try to find technical solutions. In this chapter we will discuss some of these daring ideas and projects.

© Springer International Publishing Switzerland 2016
P. Linde, *The Hunt for Alien Life*, Astronomers' Universe,
DOI 10.1007/978-3-319-24118-0_12

Psychology and Politics

The enormous distances present the fundamental challenge. Evidently, such a trip will call for a spacecraft with completely different specifications than currently available. In an earlier chapter we examined the few spacecraft that are in the process of leaving for the stars. A simple calculation shows that a spacecraft with a speed like that of Voyager 1 would take at least 75,000 years to reach Proxima Centauri. This is because Voyager is only moving with a speed corresponding to about 0.006 % of the speed of light. We need vehicles that can achieve much more substantial fractions of the speed of light.

The problem, however, is not only of a technical nature. Clearly an economic effort of colossal proportions is necessary. But a political driving force is also needed, derived from a positive public attitude on a global scale. At the same time, we face serious psychological barriers. In the current scientific community it is not unusual for a large project to take a couple of decades between initial conception and final implementation. Such timelines accompanied the spaceflight to Pluto with the New Horizons spacecraft, the construction and deployment of a space telescope like the Hubble or a giant telescope like the E-ELT. An important factor for the completion of such projects is that the individuals who begin the project have the opportunity to see it through. Starting projects with a duration longer than 50 years might well prove to be a barrier for an individual scientist. To stage and manage very long-term goals is usually not a favourite occupation for governments, who often prefer (and need) to fulfill goals within a mandated period. Long-term processes or projects are, however, not unique and do not need to pose an insurmountable barrier. To a large extent, it depends on what kind of compensation or reward is achievable. In medieval times, it was not unusual for a cathedral to need a century of construction time. On an individual level, people are often likely to save for their retirement, decades in advance. A global project that has some resemblance to a future interstellar journey is the project to balance the climate of the Earth. Here, enormous resources, great political driving force and a considerable long-term perspective are needed. It is obvious that an implementation of an interstellar

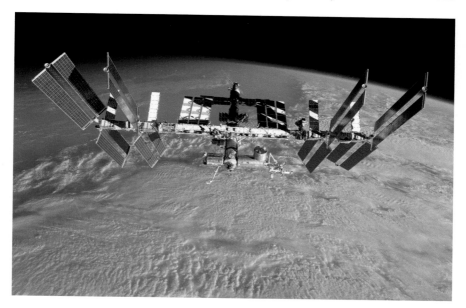

FIG. 12.1 The International Space Station, ISS, an example of a large international cooperation (NASA)

spacecraft project would require momentous efforts to obtain public support and enthusiasm, as well as inspirational political leadership. Nevertheless, a project like the International Space Station (ISS) is a promising example of international large-scale space collaboration (see Fig. 12.1).

On September 12 1963, president Kennedy made the famous statement: "We choose to go to the Moon within this decade … not because it is easy, but because it is difficult." The spirit behind this goal, on a global scale and with the words "Moon" and "decade" changed into "stars" and "century" is what would be required. However, the world does not seem to be ready for such a commitment. For the time being, poverty elimination and climate control obviously lie much higher on the agenda. A fortunate solution to these more immediate and urgent problems would give an indication of humanity's capability to handle and solve major global projects. If these problems could be solved, say within a couple of decades, a bid for an interstellar journey may then become more realistic.

Physical Limitations

Unfortunately, the greatest obstacles for interstellar space travel are determined simply by the laws of physics, at least as we perceive them. From the theories of Albert Einstein we learn that an object with mass cannot reach or exceed the speed of light. From the special relativity theory, many are familiar with the fact that time in a system moving close to the speed of light progresses slower than for an external observer at rest. For example, at 87 % of the speed of light, time progresses at half speed, and at 99 % it progresses at only one seventh the rate. This property would actually be working in favour of interstellar travel, if such speeds could be attained. More problematic is the fact that the mass of an object increases as you approach the speed of light. To get a spacecraft with a mass of several tons to accelerate to near light-speed is, in itself, a great problem, and the effects of relativity make the problem even worse. However, for most of the projects we will discuss here the velocities do not enter into the domain where relativistic effects play a major role.

Anyway, according to accepted physics, a spacecraft can never reach velocities in excess of the speed of light. This implies that even journeys to the nearest stars will take several years, in practice perhaps decades. But we can still allow ourselves to imagine that the physics of tomorrow will be more forgiving in terms of long space journeys.

Several attempts have been made to compute the energy consumption necessary for an interstellar voyage. This would depend on the mass of the spacecraft, the speed to reach and how much time is available for the trip. If the spacecraft must also be braking at the target and perhaps finally return to Earth, the energy consumption necessary becomes even more extreme. Different estimations show different figures, but many of them agree on an energy consumption comparable to or exceeding the entire current energy production on Earth. The economic investment needed for such a venture is truly enormous.

It thus appears that the conditions are lacking to initiate an interstellar journey within the foreseeable future. This, however, is not stopping enthusiasts of various shades, some at a very serious level, from studying the technological assumptions to try to find realistic solutions.

Candidate Stars

As we have emphasised, it is a long distance to the closest star. But can we say something more definite about the stars in the solar neighbourhood? Are there any exciting destinations? In view of the continuous discoveries of new exoplanets the picture is getting clearer.

We start by limiting realistic destinations to stars within a distance of 15 light years. Within this radius, it can be reasonably estimated that a journey could be completed within 100 years. We find that inside this sphere there are about 50 stars. There could be more, but in that case they would be too faint to be of interest to us. Out of the approximately 50 there are nine that can be seen by naked eye in the sky. The most well-known (and brightest) are Sirius, Procyon and α Centauri (only visible in the Southern sky). Two of them are solar-type stars (α Centauri A and τ Ceti) while about 40 are red dwarfs. As we know, these stars are smaller and cooler than the Sun and have a much longer lifetime. Recent research indicates, that such stars may well have planets at suitable distances from their parent star. At the same time, it is unclear whether these planets are suitable for life.

The first interstellar journey will most probably be unmanned, and the spacecraft will be dispatched for scientific reasons. What could be the scientific arguments to initiate such a journey?

- Studies en route of the interstellar medium, i.e. conditions in between the stars, and other astrophysical observations.
- Detailed studies of the actual destination, the target star system.
- Studies of planets in the target system, including moons and minor planets.
- Astrobiological studies of some habitable (or inhabited) planet or moon in the visited planetary system.

We note that the items listed are given in reverse order of priority. The last item would impact and revolutionise our thinking most of all. The choice of destination clearly needs to be decided from knowledge prior to launch. At the end of this chapter we will return to the question of motivation and policies related to the pursuit of a project of this unique kind.

It turns out that in spite of our momentous successes in finding new planets, our knowledge of our nearby surroundings is still embarrassingly incomplete. Out of the nearest stars we currently only know of three that have verified planets. This is all. One is ε Eridani (10.5 light years), another is GJ 674 (14.8 light years). None of the planets in question are very Earth-like, nor do they lie in a habitable zone. Further negative information indicates that several of the nearest stars lack giant planets. But we should not be discouraged by this. Available statistics, based on observations made so far, show that chances are good that we will find more planets within reach, not least planets similar to Earth. As we know, these are the hardest to discover.

Considering that the exploration of exoplanets is still in its infancy, and that a journey to the stars is far in the future, we can feel confident that all facts will be available when the choice of destination must be made. Currently, several factors favour α Centauri. It is close to Earth and most reminiscent of the Sun. Components B and Proxima should be examined during the same journey. Excitingly enough, in 2012 observations verified that there is at least one planet in this system. Although that planet is not habitable, it still inspires hope that such planets may yet be found in the same system.

Propulsion Systems

Let us now have a look at the propulsion systems available to us, not just within accepted physics, but also within a reasonable framework of possibilities. Two fundamental types can be identified: (a) craft that carry their own fuel and (b) craft that get their propulsion from Earth or pick it up during the flight. In the first category we find conventional chemical rocket propulsion of the type we know so well from current spaceflights. All manned spaceflights to the Moon, unmanned probes to the solar system and transports to the International Space Station use this type of propulsion.

Chemical Rocket Propulsion

Rocket propulsion follows Newton's third law, which simply states that any directed force has a counterforce in the opposite direction. The same principle operates in pyrotechnical skyrockets as in a space shuttle. In a rocket, hot gases flow backwards, pushing the rocket forward. The so called booster rockets used in space shuttle launches actually had a solid fuel quite similar to that used in pyrotechnics. The space shuttle itself was driven by a mixture of liquid oxygen and liquid hydrogen, carried in a large tank mounted below the shuttle (see Fig. 12.2).

The chemical rocket propulsion has the advantage that it creates a large thrust, which is necessary to lift the craft and cargo out of the gravitational field of Earth. On the other hand, it is quite ineffective, as large amounts of fuel are consumed during the short period when the rocket engine is operating. For example, the two booster rockets of the space shuttle burn out in just two minutes. The two million litre capacity outer fuel tank of the shuttle is

Fig. 12.2 Take-off for the space shuttle Atlantis. The two booster rockets contains solid fuel, while the red tank contains liquid fuel to the engines of the actual shuttle (NASA)

ejected when empty, after about nine minutes. The maximum payload is about 25 tonnes, which only constitutes about 1 % of the launch weight.

Hence, chemical rocket propulsion is useful for launching cargo from Earth, but at great cost. It is also only sufficient for exploring trips in our local planetary system. But for interstellar journeys it is not an alternative.

Electrical Rocket Propulsion

Fortunately, there are more effective forms of rocket propulsion. What we fundamentally need is to push particles (reaction mass) backwards in order to make the craft move forwards. Another way to do this is to use electrical rocket propulsion, also known as ion propulsion. The laws of physics also tell us that electrically charged particles can be accelerated in a magnetic field. With powerful magnetic fields you can reach very high velocities with electrically charged ions. The main advantage is that the carried reaction mass is very small in comparison to chemical rocket propulsion. Typically, it consists of a heavy and easily ionised atom, for example the noble gases xenon or argon or the heavy metal bismuth. To achieve the necessary ion acceleration you need very strong electro-magnetic fields. Considerable energy is needed to achieve this, but in practice you are limited to energy from accompanying solar panels. An ion engine typically has a very high exhaust speed but creates very little thrust. On the other hand, an engine of this type can be operated for a very long time, building up a high velocity for the space probe.

There are several experimental engines of this type and some have already been used in space. The European SMART-1 satellite (see Fig. 12.3) was developed in Sweden and launched in 2003 aboard a chemical Ariane rocket. After entering orbit, the ion engine was ignited. It created a meagre 0.00002 g of acceleration, compared to the approximately 3 g of the space shuttle. Then again, it could be kept running for almost 4000 hours. Consequently, after about 13 months of travel it arrived in lunar orbit and could initiate its planned investigations.

So, the ion engine concept is very promising. But the problem for larger craft and higher speeds is that much energy is needed

Fɪɢ. 12.3 The European SMART-1 space craft with an ion engine drive (ESA)

for the ion acceleration. Solar energy may be enough for slow interplanetary trips, but it is certainly no alternative for interstellar journeys.

Sailing in Space

Intriguingly, it is possible to use almost the same sailing techniques in space as have been used for centuries by seafarers. Instead of a blowing wind, in the solar system you can use either the sunlight itself or the so called solar wind. The sunlight comes in the form of massless photons, but it can nonetheless create an impulse which manifests as a pressure against a surface. However, this pressure is quite small. At the Earth's distance from the Sun, a solar sail with a reflective surface of a square kilometre would be subjected to a total pressure of 9 Newton. That corresponds to the weight of a single kilogram on Earth. If the sunlight is to be used for propulsion, extremely large and light structures are needed. Closer to the Sun, say at a tenth of Earth distance, the force is about 100 times larger, while in the outer parts of the solar system it is more or less negligible.

Nevertheless, the first experiments with a solar sailing craft have already been done. The Japanese solar sailing spacecraft IKAROS (Interplanetary Kite-craft Accelerated by Radiation Of the Sun) was launched in 2010 together with another spacecraft bound for Venus (Fig. 12.4). From its name, we realise that the craft was more similar to a kite than to a sailing ship. After a while, IKAROS deployed its 200 m^2 sail, attaining a weak acceleration.

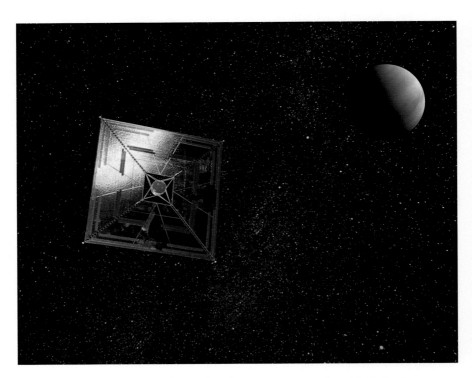

FIG. 12.4 The Japanese solar-sailer Ikaros was tested in 2010 during a voyage to Venus (Andrej Mirecki/Wikimedia Commons)

After six months, the craft had increased its velocity to about 100 m/s. Three years later, the speed was up to 400 m/s. By varying the reflectivity of the outer parts of the sail it also proved feasible to achieve some manoeuvrability of the craft.

Japan has plans for a more ambitious journey to Jupiter around 2020. Solar sailing will then be combined with an ion engine. This solution has the advantage that the solar sail can also double as panels for the necessary electrical power. The plan is to first dive close to the Sun in order to utilise the high light intensity to achieve a reasonable initial velocity, and then to accelerate further by means of the ion engine.

However, the prospects of using solar sailing as propulsion power for a journey to the stars appears very bleak. The intensity of the sunlight falls off with the square of the increasing distance. But, as we will soon see, engineers have some tricks up their sleeves that can improve the situation.

Fission and Fusion Propulsion

There is no avoiding the fact that interstellar travel will need power sources continuously generating enormous amounts of energy for many years. Within the framework of present physics, nuclear power first springs to mind. Nuclear power has already been used as an energy source (battery) for interplanetary spacecraft in the form of radioactive isotopes. Plutonium-238, with a half-life of 88 years, serves as fuel. This radioactive substance heats one end of a so called thermoelectric element, which in turn generates electricity. This type of battery has enabled the long duration of the Voyager spaceflights. They are estimated to have enough power for communication until about the year 2020. So far, nuclear power has not been used for direct propulsion. But a number of projects have been discussed and planned to some detail.

Project Orion

During the 1950s and 1960s the US defense agency ARPA analysed the potential for using nuclear energy for space travel. The project, named Project Orion, was top secret. Today the secrecy has been lifted and the documents reveal ideas that were revolutionary and exciting, but also possibly unrealistic and illusionary. The civilian successor to ARPA, NASA, was founded in 1958 and a report on nuclear powered propulsion was published as late as 1964. The report makes for fascinating reading. The project had been developed by General Atomics, then a division in the defence-oriented giant company General Dynamics.

The first ideas were sketched in the 1940s, just after the incredible inherent energy (and devastating effects) of atomic energy had been unleashed, contributing to end the second world war. Perhaps naively, at the time many considered nuclear energy as a major contributor to the bright future of mankind—even applied to spaceflight.

In 1957 an incident occurred during the testing of a nuclear device which illustrates the possibilities and the dangers. A small bomb was exploded in an underground shaft. The top was sealed with a 900 kg heavy steel lid. The bomb was more powerful than expected. A high speed camera caught a single frame of the blown-away lid, which was never seen again. It was estimated to have reached a velocity in excess of 60 km/s and may simply have been vaporised against the atmosphere, like a meteor in reverse! Alternatively, it may have become the first man-made object in space.

The planned "atomic rocket" of Project Orion (Fig. 12.5) was to be propelled by a series of small nuclear explosions, detonated behind the craft at a speed of about one per second. The technique was known as nuclear pulse propulsion. The charges would be small, less than 1 kt (the Hiroshima bomb was 15 kt). A massive rear plate would take part of the impact and in turn propel the craft forward. In order to avoid smashing cargo and crew, a two-stage shock absorber would cushion and redistribute the power (see Fig. 12.6). In this way, the maximum g-force would never exceed 3 g. Two different designs were discussed, one with a diameter of 10 m. This would fit atop the giant Saturn V rocket, having the same diameter, which simultaneously was under development by NASA. The second design was colossal-size, with a diameter of 20 m and a launching weight of about 6000 tonnes. The radiation exposures were not considered to be unacceptable. The maximum allowable radiation dose was set to 500 millisieverts, about ten times higher than today's recommendation for personnel working

FIG. 12.5 A conceivable design for the atom bomb-driven Orion spacecraft (NASA)

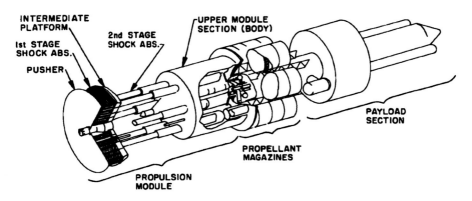

FIG. 12.6 An illustration from the NASA report on the Orion project. The sketch shows the major features of the planned nuclear-driven vehicle (NASA)

in radioactive environments. Atmospheric contamination was an unsolved problem, but in later versions of the project the intention was to have the nuclear drive take off from orbit, not from the ground. In fact, some technical experiments were performed in Project Orion, although not with nuclear explosions involved.

Spaceflight at this time was young and the space race had just begun. The visions of Project Orion were by no means modest. They were aiming for a possible trip to Mars in 1975 and a journey to the Jupiter moon Callisto in 1980, with a crew of 20–50 people! In general, the report from 1964 can be described as quite optimistic and well thought through. It advocated a number of potential advantages with the design.

Project Orion was primarily intended for travel within our solar system. Ideas were nevertheless put forward for a "super-Orion" craft that could be used for an interstellar flight, reaching a top velocity of about 5–10 % of the speed of light. A fascinating aspect of this project was its ambition to achieve all this with the technology of the 1960s. However, mankind's problematic relationship with nuclear power usage necessitated shelving the project. In 1963, the USA signed the treaty on banning nuclear tests on the ground and in the atmosphere. This is still in effect. Thus the possibilities of further advancing the project disappeared. Instead the efforts focused on the chemical propulsion rockets that led to the moon landing in 1969.

Project Daedalus

The British Interplanetary Society (BIS) was founded in 1933. With their motto, "From Imagination to Reality", this organisation would eventually become an important centre for studies of future spaceflight, standing as a source of inspiration for many scientists. The Society's many renowned members include the famous science fiction author Arthur C. Clarke. A number of original and creative ideas have seen the light of day through this organisation. But by far their most well-known study goes under the name of Project Daedalus.

In 1973 the society convened a meeting to discuss guidelines for a very ambitious spacecraft, with a capacity to reach the closest stars within 50 years. Project Orion had, in its way, been groundbreaking by describing a nuclear powered spacecraft that, using the technology of the 1960s, had a reasonable chance to actually reach the stars. The BIS intended to build on these ideas and further increase the understanding for—and the realism in—such an undertaking. The prerequisite now was that contemporary technology would be used, with some room allowed for future improvements. A group consisting of thirteen scientists and engineers sat down and started to plan a new design. Five years later, the group presented their result in a classical report that today constitutes a standard reference for all discussion on interstellar spaceflight.

The Daedalus spacecraft extends the Orion concept, inheriting the idea of nuclear pulse propulsion. But instead of using the rather crude atomic bomb propulsion, fusion propulsion was recommended, using the same power source as the stars (and hydrogen bombs). In fusion, light atomic nuclei (e.g. hydrogen or helium) combine together, while fission processes use the principle of splitting heavy atomic nuclei, for example uranium. Fusion produces considerably higher energy yields than fission. In the Daedalus rocket engine, fusion would be initiated by targeting very powerful electron beams on small pellets containing deuterium and helium-3. These microscopic hydrogen-bomb explosions would be repeated 250 times a second, while a powerful magnetic field would direct the resulting energy backwards in a jet beam with an exhaust velocity of about 10,000 km/s. The thrust would be approximately equal to that of the space shuttle but with the important difference that the engine could run for years. In the case of the Daedalus spacecraft, the resulting acceleration would be about 1/100 g.

There are many problems associated with interstellar space travel. For example, the space in between the stars is not empty. Thin hydrogen gas mixes with occasional traces of small solid dust particles. The solar system is currently passing through the "local interstellar cloud", which is assumed to have a density corresponding to one atom per every third cubic centimetre. Even though this is an extremely thin gas, it still complicates the passage for a craft that is passing through the medium with a speed of about 36,000 km/s, which is the case for Daedalus. To protect itself the craft has two barriers. After finishing the rocket propulsion phase, Daedalus sends a cloud of a small particles ahead of the vehicle, with the function of averting most of the micro-collisions. The second barrier is a more robust protection, in the form of a beryllium shield sitting at the front of Daedalus.

Another interesting problem is whether systems and mechanisms can be made to function for many decades without malfunctioning. So far, experiences are quite good. For example, the Voyagers are still operating without major problems, after almost 40 years of travelling. However, would any system on-board the Daedalus be malfunctioning, or even destroyed, the plan is to have advanced AI-guided robots to make the necessary repairs.

The design of Daedalus is such that it cannot be launched from Earth, due both to its weight and its appearance. The plan instead is to manufacture it in Earth orbit. This necessitates some kind of pre-existing space-based infrastructure. Here, some headway has already been made in the form of the International Space Station. Still another complication is the composition of the fuel. Deuterium is an isotope of hydrogen with a neutron and a proton in the nucleus. In principle, this can be extracted from ordinary sea water. Helium-3 is an isotope of the basic element helium and is, in contrast, very rare on Earth. To access sufficient quantities requires searching for it out in the solar system. The project designers suggested extracting it from the Jupiter atmosphere over a period of 20 years. However, lifting matter out of the Jupiter gravitational field is not an easy task. Lately, an alternative solution seems more realistic, namely to extract it on the Moon, where it is supposed to exist on the surface in useful quantities.

The spacecraft is planned as an unmanned science probe and is intended to visit Barnard's star at a distance of 6 light years from Earth. In the 1970s there was reason to believe that this star had planets, something which subsequently could never be proven. Travel time was estimated at 50 years. The spacecraft would consist of two stages with a total weight of 56,000 tonnes. This can be compared to the 2000 tonnes launch weight of the space shuttle. The payload was set to 450 tonnes. The first stage would run for a year, after which the ten-times-lighter second stage would operate for another two years. The final speed is estimated at 36,000 km/s or 12 % of the speed of light. After the first 4 years of rocket propulsion the craft would then fly freely for another 46 years. No option to decelerate the probe at arrival was included in the project, so the studies of the star and its planetary system would have to be implemented in a very short time. To achieve this, the payload included two 5 m optical telescopes, two 20 m radio telescopes and 18 specially equipped smaller space probes. Well in advance the telescopes would explore the system, and the probes would be launched towards planets a few years before arrival. Using ion propulsion, they would perform the necessary course adjustments. Because of the flight speed, the probes would only have a few seconds available for the most detailed studies of the planets.

Communications with the escaping Daedalus is in itself a problem. Here again the laws of physics provide obstacles. First, the time delay becomes an aggravating problem with increasing distance. At the target, Barnard's star, the delay for receiving or sending signals would be a full 6 years. In addition, any form of communication over such distances becomes extremely diluted. The radio link would be maintained using the second stage 40 m diameter exhaust nozzle as a communication dish. But the physics about diffraction of electromagnetic waves sets limits for the maximum possible of focusing of a radio beam. In order to receive a signal carrying information from Daedalus, a very large receiving system would have to be set up on Earth. However, a later study showed that laser-based communication would be advantageous, although it would require higher precision in the pointing.

The entire visit in the visited stellar system must, in any circumstances, be very well prepared in advance. On site autonomous systems with artificial intelligence must be capable of making necessary decisions and controlling the actions.

Icarus: Son of Daedalus

In 2009, 30 years after the publication of the original report, the BIS decided to revisit the Daedalus project. In the light of the fast developments in technology, members felt that many improvements could be identified. The result was a new study called Project Icarus. This study was begun in collaboration with the American organisation Tau Zero, which has a similar objective. In 2011 the project initiated a new organisation, the Icarus Interstellar. The study is well under way with a report soon to be published.

Since 1978 a lot has happened in the fields of technology and science that Daedalus depends on. Much progress has been made in the field of fusion research, and the concept of pulsed nuclear fusion has actually been tested. But much work remains before fusion energy is tamed and available for energy production in a controlled way. Scientifically our view of exoplanets has progressed from speculation to reality, as we have discussed at length in Chap. 6. Within information technology, advances are still more or less following Moore's law. The extrapolated ideas from 1978 concerning advanced computer control in the form of artificial intelligence now seem much more realistic. Additionally, end-to-end simulations of the planned functions and systems of the spacecraft can now be performed, in a way that was impossible in 1978. To facilitate communications an idea to use the Sun as a gravitational lens has been presented. We now know that the Sun measurably bends space-time in its proximity, exactly as Einstein had predicted. In the case of the Sun, the effect is quite small, but at a distance of approximately 550 AU (1 AU is the Earth's distance to the Sun) there exists a "focus" for this gravitational lens. To place the Daedalus receiving station in this point would be very advantageous.

Additionally, modifications to the expedition itself will become necessary. Apart from changing the actual target of the expedition, ways of decelerating the entire spacecraft, or at least parts of it, are also under discussion. This would enable a more detailed exploration of the target system, including soft landings on planets.

A fascinating idea with actual plans for implementation is the Virtual Icarus. Since the new studies will demand a large number of computer simulations of various on-board systems on Icarus, the idea is to link all these together to get a full simulation of the entire craft and its space-flight. The general public would then be able to follow this fictitious voyage via a specially developed web site. In this way, a detailed understanding of the functions and problems would be obtained. In addition to the obvious PR value, the ability to follow such a colossal virtual expedition in real time ought to generate much interest, not least among young people. New talents could then help to keep the project alive for a long time.

The Bussard Ram Jet

Complicating the type of rocket propulsion used in the Orion and Daedalus projects is the need to bring along enormous quantities of fuel. The weight of this fuel always dominates the total launch weight of the vehicle. In the case of Daedalus, 89 % of the original 56,000 t is pure fuel weight. Imagine if it would be possible to develop a similar thrust without actually carrying the fuel! It sounds impossible but is, at least in principle, possible to achieve.

As we have mentioned earlier, thin hydrogen gas drifts between stars, forming part of the interstellar medium. In 1969, the physicist Robert W. Bussard suggested building a spacecraft that would use this thin gas as fuel (Fig. 12.7). The principle behind the idea takes advantage of a concept known as a ram jet. The interstellar hydrogen would be collected at the front of the craft, using a huge scoop. The scoop would not be made of solid matter but consist of a force field, either electrostatic or magnetic,

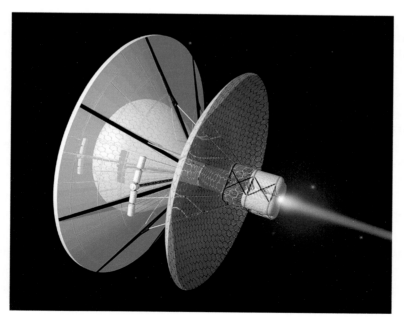

FIG. 12.7 A proposed design for a ram jet craft according the Bussard's principle. At the front magnetic field is created which collects the interstellar hydrogen (NASA)

with a size of several kilometres. The hydrogen would then be directed toward a chamber where it would be fused under high pressure and temperature. A thrust would be formed by a backwards-directed beam. Acceleration would be maintained over a very long period and the vehicle would thus be capable of reaching a speed approaching that of the speed of light.

A closer look at this ideal spacecraft, however, reveals major difficulties, The interstellar hydrogen gas in the neighbourhood of the Sun is probably even thinner than assumed by Bussard, and it is also unclear to what degree it is ionised. Only charged particles can be affected and redirected by a force field. Another big problem is that the collected ionised hydrogen, which simply corresponds to protons, only has a small tendency to ignite into a fusion reaction. Even at the centre of the Sun, where fusion processes run continuously, a single proton has good opportunities to survive at least five billion years before it will be fused with another proton. Yet another complication for the Bussard concept is the fact that the gas itself constitutes an obstacle, like a headwind for a cyclist.

But shame on him who gives up. So a number of improvements of the original design have been put forward. The craft itself could, for instance, carry coal. Not for a steam engine(!), but in order to permit a different and must faster nuclear reaction chain, the so called CNO cycle. Further proposals include using the accumulated hydrogen only as reaction mass in a craft with its own energy source. An even more radical improvement calls for an electromagnetic cannon to eject fuel in advance, in the form of deuterium, along the planned trajectory of the voyage.

Laser-Powered Sailing Crafts

As we have seen, another way to travel without carrying fuel is space sailing. To sail in space is entirely possible, as long as you are doing it in the neighbourhood of the Sun. In 1984 physicist Robert L. Forward presented an idea which extended the sailing concept into interstellar journeys. Forward was also a renowned science fiction author which, by the way, is not uncommon among physicists and astronomers. In the novel *Rocheworld*, published in

1990, he fully utilised the concept to describe a voyage to Barnard's star. He postulated a sailing craft that would be illuminated by a giant laser. The light pressure would be capable of accelerating the craft at 0.01 g far outside the solar system.

With a payload of 3500 tonnes (and a crew of 20 people) the craft had a reflecting sail made of very thin aluminium foil, with a diameter of 1000 km. This was illuminated by a battery of 1000 gigantic laser cannons, with a total output of 1300 TW. Locating them in orbit around Mercury would allow tapping solar power for running the lasers. At Mercury's distance from the Sun the solar energy is five times higher than in Earth orbit. To further focus the laser light in the direction of the craft, a specially designed lens with a diameter of 100 km was to be situated some distance out in our solar system.

A nice feature in Forward's design was the ability to also use the laser power for decelerating the craft upon approach of the target. A smaller part of the craft, including the central part of the reflecting sail, was detachable and turned around in the direction of motion. As before, the outer part is hit by the laser light and continues its acceleration. But now it reflects the light back onto the smaller 300 km sized sail carrying the payload. This is, therefore, successively decelerated and eventually can enter into an orbit around the target star. Figure 12.8 shows Forward's laser sail principle schematically.

Forward's basic idea has since been followed up and refined by several studies. The spacecraft in Forward's science fiction novel weighed in at 82,000 tonnes, the sail alone weighed 71,500 tonnes. An extremely light sail material is clearly essential. Another problem is that the sail must endure a high temperature since the power from the laser light can be several thousand times stronger than the sunlight at Earth (1300 W/m²). Later research has shown that the sail could be made of a 57 nm layers of dielectrical zirconium oxide or aluminium oxide. A nanometre is a millionth of a millimetre, so it is an extremely thin sail. These materials additionally have a very high reflectivity and can survive higher temperatures. Calculations show that this would allow for even higher laser power, which can be translated into obtaining the same acceleration with a smaller sail surface.

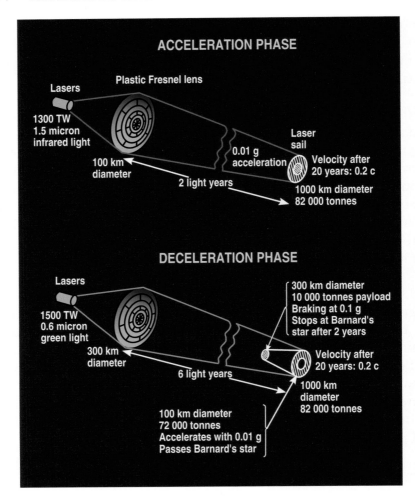

ACCELERATION PHASE

Plastic Fresnel lens

Lasers

1300 TW
1.5 micron
infrared light

100 km
diameter

2 light years

0.01 g
acceleration

Laser
sail

Velocity after
20 years: 0.2 c

1000 km diameter
82 000 tonnes

DECELERATION PHASE

Lasers

1500 TW
0.6 micron
green light

300 km
diameter

6 light years

300 km diameter
10 000 tonnes payload
Braking at 0.1 g
Stops at Barnard's
star after 2 years

Velocity after
20 years: 0.2 c

1000 km
diameter
82 000 tonnes

100 km diameter
72 000 tonnes
Accelerates with 0.01 g
Passes Barnard's star

FIG. 12.8 In the science fiction novel *Rocheworld* Robert L. Forward describes how a sailing spacecraft can travel to another star (Peter Linde)

More Exotic Methods

In Table 12.1 we compare a number of different propulsion techniques. All these operate (more or less) within accepted physics even though some of them may never be realised.

Antimatter is considered the ultimate rocket fuel. By combining antimatter with normal matter the sum of the matter is converted into pure energy. Although large parts of this cannot be

TABLE 12.1 Comparison between different propulsion methods

Method	Exhaust velocity (km/s)	Thrust (N)	Firing duration	Maximum delta-v (km/s)
Solid-fuel rocket	<2.5	$<10^7$	Minutes	~7
Liquid-fuel rocket	<4.4	$<10^7$	Minutes	~9
Ion thruster	15–210	10^{-3}–10	Months/years	>100
Solar sails	299792:Light	9/km^2 at 1 AU	Indefinite	>40
	145–750:Wind	230/km^2 at 0.2 AU		
		10^{-10}/km^2 at 4 ly		
Nuclear thermal rocket	9	10^7	Minutes	>~20
Magnetoplasma Rocket (VASIMR)	10–300	40–1200	Days–months	>100
Orion Project, atom bomb-driven	20–100	10^9–10^{12}	Several days	~30–60
Space elevator	N/A	N/A	Indefinite	>12
Project Daedalus, nuclear pulse-driven	20–1000	10^9–10^{12}	Years	~15,000
Fission-fragment rocket	15,000	?	?	?
Fusion rocket	100–1000	?	?	?
Antimatter catalyzed nuclear pulse-driven	200–4000	?	Days–weeks	?
Antimatter rocket	10,000–100,000	?	?	?
Bussard ramjet	2.2–20,000	?	Indefinite	~30,000

used for rocket propulsion (for instance the neutrinos created), it would nevertheless be widely superior to any other form of fuel, one hundred times more efficient than fusion-based propulsion. We know that antimatter exists, but presently it can only be produced in microscopic amounts and at huge costs. The particle accelerator at CERN can presently produce about 10 million antiprotons per minute. At this rate it will take about 100 billion years

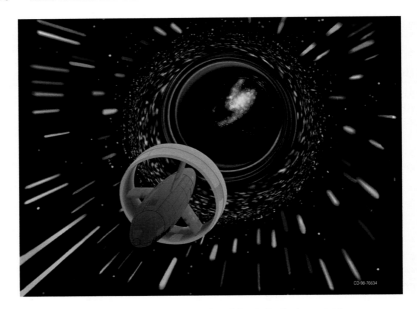

Fig. 12.9 A fictitious journey through a black hole (NASA)

to produce one gram of antihydrogen. Nonetheless, antimatter is seen as a promising future possibility, at least in combination with pulsed nuclear fusion propulsion.

In the science fiction literature it is usually a necessity to comfortably travel and communicate in the universe far faster than allowed by the speed of light. Many have followed the adventures of the Star Trek Enterprise spaceship in its exploration of our galaxy. Enterprise utilises a warp drive to quickly bridge many light years in a short time (see Fig. 12.9). Is there any realism at all behind this notion?

We know today that three-dimensional space is curved in the presence of gravitational fields. It is also acknowledged that exceptionally intense gravitational fields are created by very large and dense masses. The most extreme examples are the black holes. In this case, the curvature of space-time becomes very large, and some speculate that a black hole could "short-cut" two very distant points in the universe. In such a case a further speculation might be to attempt to travel through this hole, arriving at the other very distant point. If this were possible, it might provide a means of circumventing the limit of the speed of light.

Another speculation that sometimes is put forward, is that the spacecraft itself would be capable of affecting the space in such a way that space-time in front of the craft is compressed and the space-time behind it is expanded. The spacecraft would then be moving inside a "bubble" with the effect that enormous distances could be bridged in a short time. This speculative propulsion method would demand enormous amounts of energy, and also the existence of negative or so called exotic mass, both of which are currently only hypothetical concepts.

Naturally we know very little about what possibilities future physics may give us. One might just contemplate how much Alfred Nobel would have understood about the discoveries that his prize annually is rewarding. In modern physics new concepts like vacuum energy, quantum fluctuations, eleven-dimensional string theories, repellent dark energy, etc, are commonplace. Perhaps the seeds for more practical interstellar travel may lay within this world at the limit of current knowledge.

Why Make the Journey?

Now let us return to more fundamental and philosophical questions concerning a project like a trip to the stars. Is it really a sensible project? Can it ever be possible to get the necessary support for an effort which demands that type of gigantic resources? Is it not just a romantic dream amongst science fiction nerds or derailed scientists?

In 2011 the US Defense agency DARPA, together with NASA, decided to look into the matter. The study is called The 100 Year Starship Study (100YSS). A smaller conference convened at the beginning of the year with a number of experts invited. On 11/11 2011 this study concluded with a major conference in Orlando, Florida, USA. The main motivations for interstellar travel were summarised in the following way:

- Human survival. We need a "backup" of the human culture and the biosphere of the Earth, in case of a global calamity (for instance a collision with a large asteroid).
- Contact with other life. Is there life elsewhere, even intelligent life? Or are we actually alone in the universe?

- Evolution of the human species. Expansion of the human understanding and awareness, through space research.
- Scientific discoveries. Breakthroughs in the natural science understanding of the universe, a hunt for new knowledge.
- Conviction and faith. The search for God and the divine, a need to explore beyond the Earth's atmosphere as a natural part of theology or through religious revelations.

The last point undeniably suggests the width among the experts present. Also clear was that the human element must be present very early in the attempts to reach the stars. Here we have explicit experiences to learn from. During the 1960s and 1970s it was the manned spaceflights that caught the interest of the general public. When Neil Armstrong first set foot on the Moon, it was a global event seen all over the world. And when the Apollo 13 crew was near disaster the interest was also very high.

But how many know that in 1970 the Soviet Union succeeded in a soft landing on the Moon and, using an automatic vehicle, dug up 100 g of lunar matter successfully returning it to Earth? Since 1972 no human have visited the Moon. A series of extremely successful interplanetary expeditions were carried out in the 1970s and 1980s, to the outer parts of our solar system (see Chap. 8). But all were unmanned and, even though science was hugely enriched and scientists celebrated, the public was rather lukewarm in their overall response.

It is quite obvious that in order to capture the necessary interest and support for a trip to the stars, humans of flesh and blood must be deeply involved. At the same time it is clear that some unmanned probe must come first. This is also a lesson from all space research.

How Do You Create a Realistic Project?

100YSS also discusses organisational aspects of a journey to the stars. How and by whom should a project of this magnitude, with an extreme long-term vision, be pursued? Should it be governmental or private investments? Or could the backbone be a popular movement? One example, mentioned in this context, was the Nobel foundation, which has now existed for over 100 years. The Nobel prize constitutes the ultimate scientific acknowledgement and has in many ways, both directly and indirectly, stimulated research and innovation. With a comparatively small investment coupled to the right methods, great things can be achieved. Fantasy and creativity need to be stimulated and encouraged. 100YSS gives a few suggestions of how this could be implemented:

- In a project running for a hundred years you need to define milestones, each one creates an interest, new energy and new will of investment. Examples of such milestones could be manned expeditions to Mars and Jupiter and then to Oort's Cloud of comets at the outer edges of the solar system. In fact the Moon-landings were prepared and built up along these lines.
- Irrefutable evidence of other planets similar to Earth must be presented. A clear and attractive target will increase the interest in an interstellar expedition.
- Science education must be strengthened at all levels.
- The marketing must be very effective. Credibility for the project must be established through engaging well-known personalities that support it. In a collaboration with the film industry a major movie, based on the ideas of the projects, could be produced in order to catch the interest of the public on a large scale.
- It is crucial to secure a strong scientific leadership, where members are prepared to take risks. In the information technology industry interesting examples as role models are Steve Jobs (Apple) and Bill Gates (Microsoft), both very successful.
- You have to create alternative funding methods, aiming both for large donations (compare the large Keck Observatory, financed by a donation of $140 million) and for millions of micro-donations from the public.

This Century—Or the Next?

A study of the miscellaneous suggestions for interstellar expeditions that have been put forward in the latest decades shows that travelling to other stars is very difficult but fully possible. One conclusion is that we are now in between different phases in the development of space flight, where the political interest is minimal and the commercial driving forces still are in their infancy. It can be seen from the studies made in the 1960s that optimism for the future was considerable, as was the technological potential. If the spirit of the 1960s had prevailed, we would already have had a colony on the Moon and made manned expeditions to other parts of the solar system.

But initiatives like the Project Icarus and the 100 Year Starship Study offer hope for the future. Important trends have surfaced which could not be foreseen in the 1960s, both within information technology and in extreme miniaturisation (nanotechnology). Perhaps the first interstellar spacecraft will not be a monster of thousands of tonnes at all, but instead a small high-tech probe weighing less than a kilo?

Things have also recently happened in the marketing approach. A well publicised and controversial project is Mars One. The goal is to establish a colony with 20 people on Mars no later than 2035. Starting in 2012, anybody could apply to become a participant in the one-way-ticket expedition. Organisers project that funding will come from a variety of sources, including application fees, television rights, and crowd-funding. The future of this particular project appears unclear, but it demonstrates new thinking in space flight. In addition, the recent science fiction movie, *Interstellar*, could be a precursor to raise the general interest for interstellar voyages.

Undoubtedly, the coming decades will see an expansion in our own solar system, as humanity gradually sets up colonies on the Moon with extended space stations in Earth orbit. Then, perhaps the infrastructure will be available to facilitate the really long journey. After all, deep within the human nature lies the desire to continue exploring. And now it is outwards, first towards the planets and then, ultimately, the stars where we today find the irresistible challenges…

The only question is whether the first journey to the stars will happen in this century or in the next.

13. What Do the Aliens Look Like?

To this point, we have not discovered life anywhere else in the universe. We know neither if it exists nor where it might be located. Therefore, to speculate about how aliens may look is perhaps not entirely realistic—but nevertheless quite fun. Without doubt it is a question which interests many—not least children. It is rather enlightening to see how children of today imagine extraterrestrials, for instance by asking them to draw their favourite alien. They normally respond enthusiastically, and the author has had many opportunities to inspect such colourful and imaginative drawings. Figure 13.1 shows a few examples. Clearly, the concept of aliens is quite familiar to children. They are not always humanoid in shape even if they usually exhibit some human features like eyes, ears and nose. Looking at several hundred such drawings, you get the definite impression that children in general have a rather positive view of aliens.

The same is not necessarily true for grown-ups. Clearly, the perception of aliens in the general public is strongly affected by the media's view of extraterrestrials, not least those coming from the movie industry. There, they are often portrayed as evil monsters, generally surprisingly stupid and often quite violent. There are, however, exceptions. Steven Spielberg's child-friendly movie "ET—The Extraterrestrial" from 1982 tells the story of a friendly but lost alien, and gives a positive and contrasting picture to the horror monsters from outer space. We have previously mentioned another more sophisticated portrayal in Arthur Clarke's and Stanley Kubrick's brilliant 1968 movie "2001—A Space Odyssey". It describes how an unknown and superior intelligence interferes with the evolution of life on Earth. No extraterrestrials are ever directly encountered (and therefore there is no need to show them), just traces left behind by them in the form of advanced machines. It is good science fiction, where scientific and technological correctness is combined with deeper and more

© Springer International Publishing Switzerland 2016
P. Linde, *The Hunt for Alien Life*, Astronomers' Universe,
DOI 10.1007/978-3-319-24118-0_13

Fig. 13.1 A set of children's drawings of imagined aliens. A typical feature is the fact that they seem rather friendly but otherwise exhibit only vague similarities to humans (Tycho Brahe Society)

philosophical speculations about the origin and destiny of mankind. As an aside, the film also mirrors the space optimism of the 1960s, where moon colonies and voyages to Jupiter were considered very probable for the year 2001.

Generally, it can be said that the terror created in 1938 by Orson Welles' dramatic radio theatre (see Chap. 4) could not be repeated today, as the general public (as well as science) has advanced beyond the most immediate fear of aliens. But let us begin by looking at aliens from the very beginning, i.e. at the microscopic level.

Aliens in the Microscopic World

In the chapter regarding the evolution of life, we showed why all life on Earth is based on organic compounds. The reason is that carbon is a basic element that has an especially good ability to connect to other atoms, and to itself. We also noted that water was unusually suitable as a solvent, which facilitates the transport of many important building blocks inside a cell. Naturally, we can speculate on other possible binding atoms and other solvents. In that case it might also be necessary to think in completely different terms about the environment in which a different type of life would evolve. It has been suggested that silicon-based molecule chains would survive much better in a sulphuric environment than in a water environment. If, like us, silicon-based life used oxygen for breathing, you would expect the exhalation product to contain silicon dioxide instead of carbon dioxide. However, the problem with that is the fact that silicon dioxide in principle is the same as sand, i.e. a solid substance. In an environment with sufficiently high temperatures silicon dioxide is, however, liquid.

There are also conceivable alternatives to water as a solvent. We note that is not only the temperature but also the pressure which determines when a substance can exist in liquid form. As is well known, water boils at a temperature of 100 °C, but this is valid only at normal atmospheric pressure. At a one hundredth of the atmosphere pressure, water boils at 7 °C and at a pressure one hundred times higher it will not boil until a temperature of 311 °C. We have already noted that temperatures and pressures strongly vary across the solar system. Thus, water may be liquid under rather extreme conditions, but so can other liquids. As an example, ammonia can be a liquid at low temperatures and in a rather large temperature range. The same goes for methane and ethane, which are gases at Earth conditions but liquids on Saturn's moon Titan. If a biosphere could work with such a substance as a solvent you may, due to the low temperature, expect to find the metabolism in a hypothetical life form to be very slow.

On September 29, 2010 NASA announced a press conference which, according to the pre-released statement, was "to discuss an astrobiology finding that will impact the search for evidence of extraterrestrial life". This proclamation caused great excitement within scientific circles around the world. Considering the great weight usually carried by NASA announcements, the interest in the press conference soon grew enormously. Had life been discovered on Mars? Or perhaps on one of the satellites of Jupiter or Saturn? However, when the press conference started the reaction was somewhat subdued. It turned out that the news was about earthly life—not extraterrestrial.

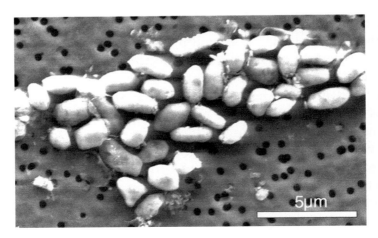

FIG. 13.2 The bacterium GFAJ-1 under a microscope. The width of the image corresponds approximately to a hundredth of a millimetre (NASA)

Still, what NASA presented was something very exciting. A team of researchers had analysed samples taken at Mono Lake, a small lake in California. Since it lacks estuaries it exhibits very special conditions; in particular the lake has very high levels of salt. One bacterium in the samples, which was prosaically named GFAJ-1 (see Fig. 13.2), turned out to have the property of surviving without phosphorus, instead utilising arsenic to grow. As we have seen in the chapter about life's evolution (Fig. 3.4), phosphorus is one of the important basic elements within the DNA framework. The scientists suspected that GFAJ-1 had replaced these phosphorus atoms with arsenic atoms. That would signify encountering a life form with considerably different DNA structure than the ones found on Earth so far. This, in turn, would increase the possibilities for finding extraterrestrial life based on environmental conditions other than those on Earth.

Unfortunately, the discovery does not seem to hold water at a closer inspection. A number of other experts immediately criticised the report, and recently several research groups have tried to repeat the results without success. So the situation right now is that also GFAJ-1 appears to have normal DNA.

The step from single-cellular life to multi-cellular life is very large—and then ever larger to intelligent life forms. On Earth, we can see how impressively life has succeeded in filling every niche with incredible variation in appearance over billions of years. But life on Earth has nevertheless kept itself within a certain framework.

It is DNA-based and the number of fundamental "construction drawings" in the animal kingdom is no more than 40, all of which actually originated during the Cambrian explosion 540 million years ago. That the DNA-molecule would be the only conceivable method of storing a genetic code is not obvious. Even with carbon-based life as a foundation, you can imagine other genetic schemes, perhaps with adaptation to environments widely different from our own.

Nonetheless, we arrive at the conclusion that the carbon-based life existing on Earth seems to be the most probable form of life. But Nature always surprises us, and to surmise something different would certainly compromise the Copernican principle. The Earth, The Sun and humanity are not located at the centre of the universe. Perhaps the same can be said for our life form?

Intelligent Aliens

Is there anything at all that we can say about the appearance of intelligent beings living on an alien planet? Well, perhaps there are a couple of things.

That life adapts to its environment is the cornerstone of the Darwinian theory of evolution. That this would also be a valid principle for alien life on other planets is not a very bold assumption. Unfortunately, this assumption does not limit the appearance of intelligent beings very much. On Earth, one species has developed awareness and dominance on the planet. Was it a coincidence or could it perhaps have been achieved by some life form other than human beings? This is, of course, open for speculation.

Certain basic characteristics for a creature, intelligent or not, are reasonably obvious, and originate from clear-cut survival factors. Our star, the Sun, radiates over a wide spectrum of wavelengths, but mostly in the yellow part of visible light. This follows fundamental physics describing the relation between temperature and light emission (see Box 6.2). That living organisms develop vision sensors (eyes) adapted for such light (or similar) is natural. Further, at least two eyes are necessary to perceive depth. The brain combines two slightly different images from the eyes to achieve a sense of depth. Admittedly, depth can also be sensed using ultrasound reflections. To perceive sound is another survival factor.

Here, too, at least two auditory organs (ears) are necessary to determine from which direction an enemy is approaching. These organs have a certain advantage by being located as high as possible, and preferably located only a small distance from the analysis and manoeuvering organ (the brain). Similar arguments can be given for other bodily functions and senses. At the same time, it is obvious that these characteristics by no means are unique to humankind. Many highly developed animals have senses clearly superior to those of man's. Others possess a few that we do not have, for instance sonar that bats and whales use for distance determination and identification of potential prey. Another example is chemical communication via extremely sensitive smelling sensors. Perhaps somewhere beings have developed natural radio communication? That would come quite close to what we call telepathy.

From a purely physical point of view, man is also not very impressive. Many animals are stronger and faster. Long distance running is an exception, however; there are few marathon runners among the animals. This property gave early man the ability to pursue and hunt prey for long distances.

Other factors evidently must have been more decisive for the rise of mankind. Our origin in a tree environment contributed to our gripping ability, which subsequently became an important factor in the development of flexible hands with fingers, not least the one with an opposable thumb. About two million years ago, when our forefathers climbed down from the trees and managed to walk on only two legs, the foundation was laid for the ability to use simple tools as weapons. The ability of throwing gave humans an advantage over other animals looking for prey. The brain was developing fast, especially the parts that handle language, judgement and social behaviour. Some scientists argue that this development was also stimulated by the change of diet as humans began cooking food on a controlled open fire. Another important social factor seems to have been the long childhood and adolescent period of man, affording rich possibilities for the learning of a language and social patterns. As mentioned earlier, we humans are thus more "software learning" oriented as opposed to "hardwire instinct" oriented. As seen from a geological perspective, the exceedingly rapid conversion from tree climbers to spacefarers is closely connected to the equally rapid development of communication and the capacity to transfer and store information.

The question we must ask ourselves is whether modern man is the ultimate and inevitable final product of the evolution on Earth. Judging from the sequence of coincidences which have affected the history of the Earth, it is difficult to believe that this would be the case. In the chapter about life's evolution on Earth we mentioned, among other things, a number of mass extinction events, whereby the latest (possibly not the last) resulted in the extermination of the dinosaurs. Evolution probably would have found other ways if these random events had not happened or happened in other ways.

Whatever happens with life on other worlds, even if these by chance would have an environment similar to Earth's, is simply impossible to predict. Not to mention predictions even harder in the very different environments that are clearly indicated among the exoplanets and their satellites. Yet, on the other hand, they may not be entirely unlike ourselves. In Fig. 13.3 we see beings that may have developed along similar lines as mankind.

FIG. 13.3 An imaginary but possible scenario, showing a scene from an advanced civilisation developed along similar lines as mankind (Mike Carroll)

The overall requirements for mobility, senses to perceive the surrounding reality, capabilities to manipulate tools and so on, are all reasonable and probable, but give ample room for speculation. What would a civilisation on a water world look like? Probably somewhat similar to our fish, but would it be possible to have an industry, even to use fire? On the other hand, the water environment may provide advantages, such as an increased buoyancy, to transport heavy structures. What would beings look like on a super-Earth, with much higher gravity than on Earth? But it is said that reality sometimes surpasses fiction. Perhaps we are speculating along completely erroneous lines?

Is There a Cosmic Evolution?

To speculate on what other beings may look like can be entertaining since we can give fantasy free rein. The drawings of the children are one example of this. As we have previously seen, most SETI researchers are of the opinion—for statistical reasons—that other civilisations would be more advanced than our own. We are then talking about a difference of hundreds of thousands, maybe millions of years. But have the speculations been drawn to their logical conclusions in terms of what such a civilisation might look like? What does happen to a civilisation that is tremendously older and more advanced than our own? How does evolution continue? Quite probably, biological evolution is overtaken by other factors.

Earlier we saw how, according to Darwinian theory, evolution has advanced life on Earth to a very high level of complexity and expansion, without even touching on topics like intelligence and civilisation. Is it possible to apply the concept of evolution into an even larger framework, for example considering the evolution of the entire universe? Some scientists believe this a reasonable scenario; they speak of the cosmic evolution. According to them, this process happens in a series of stages. The first stage is the astrophysical evolution, which started at the creation of the universe and which is still continuing. Here, we will not go into any detail of how the universe was created, even though it certainly is a fascinating story. However, the main events are discussed in Box 13.1.

Box 13.1: The Big Bang: A Summary

The universe was created 13.8 billion years ago, according to the prevailing Big Bang theory. In the initial event, all energy was created and the concepts of space and time became meaningful. An expansion of space began—and still continues—and the universe increasingly grew larger. This development progressed in rather distinct steps (Fig. 13.4). In the beginning, the universe was totally dominated by energy and radiation. Matter in the conventional sense arose after about 380,000 years, when remaining elementary particles as protons, neutrons and electrons formed hydrogen along with some helium. At this point, matter began to dominate the universe which became transparent allowing the radiation to escape. It is this radiation that we can observe today as the cosmic microwave background. The first stars probably formed about 400 million years later. It may still be possible to view the light from this stellar generation using the new giant telescopes now under construction. From then on, more stars, galaxies and

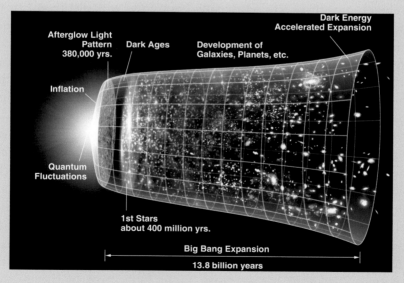

FIG. 13.4 A summary of the astrophysical development of the universe since the Big Bang. The WMAP satellite gave a detailed view of the background radiation (NASA/WMAP Science Team)

planets formed until the current epoch which we can observe all around us.

However, the matter observable today evidently only contains about 5 % of the energy content of the universe. Cosmologists assume that 27 % is so called dark matter and the remaining 68 % is known as dark energy. While there is some hope of identifying the nature of dark matter within the near future, the nature of dark energy still is almost totally unknown. It acts as a repellent force, accelerating the expansion of the universe.

The second stage is the chemical evolution, which implies a high level of self organisation where atoms are organised into more and more complex molecules. This primarily happens on planetary surfaces but also out in interstellar space. Almost 200 different molecules have been detected in space using radio telescopes. After this stage comes the biological evolution, i.e. the creation and development of life. How this happened was discussed in the beginning of this book. The last stage in this scenario—or at least the latest—is the cultural evolution. This amounts to the development of collaborating intelligences both locally on Earth and on a more universal scale, with either a proliferation of human civilisation or expansion in coordination with other existing civilisations out in the galaxy.

Accordingly, some scientists see the development of life and the growth of civilisation on Earth as a natural and almost inevitable consequence of the cosmic evolution. Using this somewhat philosophical point of view, the idea of life on other planets is also natural, as would be the existence of other civilisations. Could it be that the progress of life is a natural part of the evolution of the entire universe? That an important step was taken—at least here on Earth 3.5 billion years ago—when we entered into the biological epoch? And that we are now in the beginning of the cultural epoch? Perhaps that life is even beginning to dominate matter? It sounds—and is—speculative, but let us have a look at the underlying arguments.

The Struggle Between Order and Disorder

First and foremost, we have to discuss the struggle between order and disorder. From the physics class in school we may remember that energy gradually is converted into lower and lower states and eventually ends up as heat. This also corresponds to an increasing disorder. At the microscopic level this disorder expresses itself in various forms of motions among the molecules.

It is possible to create temporary and local order, but only at the expense of a greater disorder on a larger scale. We can create a library of books of well-organised information, but its creation demands that we expend energy, e.g. by the librarian whose work requires eating, which in turn means that other resources must be used, etc. We find a more concrete example in our refrigerators, where heat is removed and therefore order is locally increased, but only at the expense of using an electrical motor whose work is consuming energy derived from other resources, increasing the disorder on a larger scale.

So clearly the natural process is to increase disorder. How, then, can it still be possible that we see so many examples of extremely well-ordered systems, everything from galaxies to cells? Even if some are looking for answers in mysticism or religion, there is an answer well within the framework of natural science and physics. It concerns the efficiency of how energy flows and is used in an optimal way. The energy production inside the stars forms the basic cosmic engine. This energy is utilised both by nature and by biological creatures like us, in order to build small well-organised islands of order. We should, however, keep in mind that the energy of the stars eventually runs out, the stars fade. But before they do, they often eject parts of their matter, which in turn may become the embryo of new stars somewhere else, this time enriched with heavier basic elements. This really large cycle continues throughout billions of years, but by all accounts the universe, in the very distant future, assisted by an ever ongoing expansion, will end up in emptiness and darkness.

In order to create local islands of organisation it is though necessary that a system is in local non-equilibrium. This simply implies that it cannot easily interact with its surroundings since it is, in some sense, disconnected and isolated. A system in equilibrium, on the contrary, distributes its energy equally everywhere, and this suppresses the possibility for order. On the very largest scale we note that the expansion of the universe itself promotes the existence of systems in non-equilibrium: the distances increase steadily, which impairs reaching equilibrium.

The pre-condition for the creation of local islands of order is that energy is being supplied effectively and in the right amount. Too little energy starves the system, while too much destroys it. A cell is a good example of this. It contains incredibly complicated biological machinery, a complete miniature chemical plant, maintained by well-adjusted chemical energy. A modern computer could represent a corresponding electronic system, where huge amounts of work can be achieved using small quantities of energy. In Fig. 13.5 we see an attempt to illustrate how such systems—on different scales—have developed in the universe.

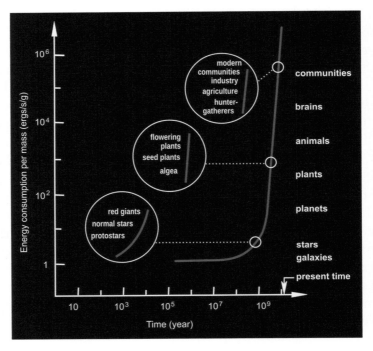

FIG. 13.5 An effort to illustrate the evolution of the universe and the Earth towards higher degree of order. The energy is progressively used more and more efficiently (Peter Linde after E. J. Chaisson)

The cosmic evolution thus is driven by the interaction between energy flows and selections. It is in this process and during long time periods, that complex systems can form.

The Cultural Evolution

The fourth stage in the cosmic evolution is supposed to be the cultural evolution. What does this concept actually mean? The culture and technology that humanity has created started to accumulate about 200,000 years ago. With the development of language, writing and the craft of printing, the cultural evolution increased its pace, and with internet as fuel the cultural evolution is now advancing at a furious speed. Our cultural level entails the sum of everything humanity has created, not just material things but even more so immaterial things like ideas, values, music, literature, etc. And above all, it concerns the communication about these things.

In fact, there is a group of scientists that are trying to understand the cultural evolution in quantitative terms. They suggest that culture is carried by memes, analogous to how life is defined by genes. A meme is often (but not always) something abstract like a word, a language, a mathematical formula or a piece of music. Memes also share another property with genes: they easily replicate themselves and are modified in the replication process. Via communication and mass media of different types, culture is conveyed and deepened. So far, the concept of memes has not been developed enough to be recognised as part of a new discipline. The intention and goal of the researchers is nevertheless to more concretely allow an understanding of the origin of a culture, and perhaps also to influence memes in a more humane direction, something that may become very important in the near future.

Is the Technological Singularity Near?

The cultural evolution has led to technological advances, which means that its biological predecessor is now out of the game. Humanity now has its destiny in its own hands; we can and will affect our future at a pace much faster than the biological evolution. The only question is in what way.

The technologies closest to affecting and determining this are probably genetic engineering, nanotechnology, space science and artificial intelligence (AI). Using genetic manipulation, scientists will be able to prolong the human life span considerably, and also create new possibilities for humans—perhaps eternal life may even be within reach earlier than many would dare to believe.

Even more likely, however, is that the cultural (and therefore the technological) evolution will make a breakthrough in artificial intelligence. There are already contemporary trends that stagger the mind. Twenty years ago, very few knew about the internet or the web. Mobile communication was more or less unknown. An important trend today is the explosive increase in capacity and proliferation of smart-phones. In recent years more than a million applications programs ("apps") have been made available, and they are rapidly approaching 200 billion downloads. So, what will happen within, say, two decades?

Ray Kurzweil is an American inventor and author with a substantial expertise in artificial intelligence. He is not afraid of making radical predictions about the future and has published much on this subject. In his book from 2005, *The Singularity is Near*, he details his predictions for the near future. Kurzweil thinks that the rapid development of computing power will continue and guesses that a computer in 2019 will pass the so called Turing test. To pass this test, answers from a computer to arbitrary questions will not be distinguishable from those of a human being, which would imply that their intelligence and awareness are also comparable. According to Kurzweil, computers thereafter become capable of improving themselves to get progressively smarter. He thinks that by the year 2045 (approximately) computers will become a new life form totally superior to mankind, by means of their superior computing power, communications fast as lightning and with access to all human knowledge. With this, the evolution would then go from biological to post-biological or, in our case, post-human. From this point forward, the development continues at a speed that might well be incomprehensible to humans. Following what is sometimes called the technological singularity, mankind would then be in danger of becoming irrelevant.

As previously mentioned, Swedish philosopher Nick Bostrom is studying the future of humanity. Bostrom points out possibilities and dangers. Like Kurzweil he discusses the possibility of a super-intelligence arising in a near future, but he approaches the topic from another angle. Bostrom notes that the characteristics and modes of operation of the human brain are understood only in broad terms so far. But this will change in the near future, and downloading the total informational content of the brain will become possible, including personalities, experiences, etc. Through very detailed slicing and thereafter scanning the structures of the human tissues, it ought to be possible to reconstruct a model mapping the neurons and their extremely complicated network. This is turn could be simulated in a super computer, with the result that you get a software-based copy of a complete human personality. This intellect could live in a virtual reality world or inhabit a robotic body in the real world. If this sounds like science fiction, it might be sobering to study the ongoing "Human Brain Project", sponsored by the European Commission. The vision of this project is formulated thus:

"Understanding the human brain is one of the greatest challenges facing twenty-first century science. If we can rise to it, we can gain profound insights into what makes us human, build revolutionary computing technologies and develop new treatments for brain disorders. Today, for the first time, modern ICT has brought these goals within reach."

If such a download becomes achievable, the super-intelligence is very near. Replication becomes easy and fast, and all that is needed is a copying operation. The new "individual" will take over all knowledge from the previous one and thus starts at a high level. The thought processes can be expected to be thousands of times faster than in a biological brain. Eternal, or at least extremely long, life will become the norm. The dependence of a specific environment decreases. For example, such an artificial life form would survive excellently in space without any major problems. Merging of the personalities, knowledge and intelligence of several individuals appears to become a realistic possibility.

Bostrom's hypothesis has at least the sympathetic quality that a future super-intelligence would have a clear foundation in human values and ideas, something which may not necessarily be true for a purely AI-based machine intelligence.

At least the younger part of the current generation will have a good chance to find out whether these scenarios will become reality. The time perspective is probably not longer than that.

Silicon and Electrons Instead of Flesh and Blood?

So perhaps the question, "What do the aliens look like?" has a surprising answer. Perhaps we will encounter not beings made from flesh and blood but instead beings made from silicon and electrical energy. Perhaps they will not even live on a planet. A mind-boggling example is outlined in the 1957 science fiction novel written by British astronomer Fred Hoyle called *The Black Cloud*. In it, Hoyle lets humanity come in contact with an enormous intelligence which exists in the form of an interstellar cloud (see Fig. 13.6).

FIG. 13.6 The dark dust cloud Barnard 68. The dust and gas obscures all light coming from background stars (ESO)

May we get in touch with such super-intelligences? Nobody can predict what they think or feel. Compared to the highly developed society in an anthill we may perhaps consider ourselves as a super-intelligence. But would anyone ever have the idea to start communicating with the anthill on their terms? The question concerning contacts with ETI has more than just technological aspects. Ethical, political and social factors play roles, as it probably will for the ETI as well. Do they want to interfere in our existence? Can we manage to survive the abyss of cultural difference that is very likely?

The theory of the cosmic evolution implies that the creation of a super-intelligence is a natural development of the universe. This may already have happened in other worlds. Or perhaps we really are the first. The future and final destiny of humanity surely is closely linked to the universe. The words of God in the Old Testament about mankind filling the Earth take on a much deeper significance in such a future perspective. Where mankind's voyage is going and how it may end, future historians will have the pleasure to tell about.

Appendix A
Some Researcher Profiles

The hunt for alien life is done by many researchers in many different disciplines. To choose a few out of many is therefore difficult. The following are examples of eminent scientists in some of the most relevant disciplines covered in this book.

Michel Mayor (born 1942) is an astronomer at the Geneva Observatory. Together with Didier Queloz he made the first historic discovery of an exoplanet orbiting a solar-type star and have since continued with more than 200 new discoveries. He uses advanced spectrographs mounted on large telescopes, using the radial velocity method to search for exoplanets. During later years, most of the observations have been made with the ESO 3.6 m telescope, at La Silla, Chile.

Geoff Marcy (born 1954) is an astronomer at the Berkeley University in California, USA. After the Mayor announcement of the first exoplanet observation, Marcy could immediately confirm the discovery. He was subsequently the first to find a multi-planet exosystem. As a leader of the Berkeley Center for Integrative Planetary Science, he has contributed to more than 250 new discoveries of exoplanets. He is active in the Kepler project and was until 2015 director for the SETI research at Berkeley University.

Carl Sagan (1936–1996) was a charismatic and sometimes controversial American astronomer, primarily active at the Cornell University in Ithaca, New York, USA. As a populariser, Sagan became famous for, among other things, his TV series "Cosmos", and the science fiction novel "Contact", which also became a movie. He played a major role in NASA's space research programme during the 1970s and 1980s. Sagan was an early advocate for SETI studies. The Carl Sagan center, which is part of the SETI institute, is devoted to research in astrobiology.

© Springer International Publishing Switzerland 2016
P. Linde, *The Hunt for Alien Life*, Astronomers' Universe,
DOI 10.1007/978-3-319-24118-0

Frank Drake (born 1930) became a radio astronomer at the Harvard University, Massachusetts, USA. He is one of the first pioneers and most important figures in SETI research. Drake's first observations, together with his well-known formula, had an decisive influence to stimulate an entire new scientific discipline. He has held a large number of leading posts in American astronomy and still is active at the SETI institute.

Jill Tarter (born 1944) is an American astronomer who has dedicated most of her life to SETI research. Already as a student, she worked on the SERENDIP project and in the 1990s she was the head of the Phoenix project, the so far most ambitious attempt to find extraterrestrial civilisations. Tarter is one of the founders of the SETI Institute and held a leading position there until 2012. She has played an important role in the creation of the Allen telescope, dedicated to search for artificial signals from space.

Nick Bostrom (originally Nicklas Boström, born 1973) is a Swedish philosopher, interested in existential matters concerning the future of humanity. He is the director of the Future of Humanity Institute and the Programme on Impacts of Future Technology, both at the University of Oxford, England. His interests include, among other things, artificial super-intelligence and future extinction scenarios. He is an advocate for trans-humanism, i.e. the opportunity for man to, through advanced technologies, enhance physical and mental capabilities.

Appendix B
List of Music on Voyagers
"Golden Record"

1. Bach, Brandenburg Concerto No. 2 in F. First Movement, Munich Bach Orchestra, Karl Richter, conductor. 4:40
2. Java, court gamelan, "Kinds of Flowers," recorded by Robert Brown. 4:43
3. Senegal, percussion, recorded by Charles Duvelle. 2:08
4. Zaire, Pygmy girls' initiation song, recorded by Colin Turnbull. 0:56
5. Australia, Aborigine songs, "Morning Star" and "Devil Bird," recorded by Sandra LeBrun Holmes. 1:26
6. Mexico, "El Cascabel," performed by Lorenzo Barcelata and the Mariachi México. 3:14
7. "Johnny B. Goode," written and performed by Chuck Berry. 2:38
8. New Guinea, men's house song, recorded by Robert MacLennan. 1:20
9. Japan, shakuhachi, "Tsuru No Sugomori" ("Crane's Nest,") performed by Goro Yamaguchi. 4:51
10. Bach, "Gavotte en rondeaux" from the Partita No. 3 in E major for Violin, performed by Arthur Grumiaux. 2:55
11. Mozart, The Magic Flute, Queen of the Night aria, no. 14. Edda Moser, soprano. Bavarian State Opera, Munich, Wolfgang Sawallisch, conductor. 2:55
12. Georgian S.S.R., chorus, "Tchakrulo," collected by Radio Moscow. 2:18
13. Peru, panpipes and drum, collected by Casa de la Cultura, Lima. 0:52
14. "Melancholy Blues," performed by Louis Armstrong and his Hot Seven. 3:05
15. Azerbaijan S.S.R., bagpipes, recorded by Radio Moscow. 2:30
16. Stravinsky, Rite of Spring, Sacrificial Dance, Columbia Symphony Orchestra, Igor Stravinsky, conductor. 4:35
17. Bach, The Well-Tempered Clavier, Book 2, Prelude and Fugue in C, No. 1. Glenn Gould, piano. 4:48
18. Beethoven, Fifth Symphony, First Movement, the Philharmonia Orchestra, Otto Klemperer, conductor. 7:20
19. Bulgaria, "Izlel je Delyo Hagdutin," sung by Valya Balkanska. 4:59
20. Navajo Indians, Night Chant, recorded by Willard Rhodes. 0:57

(continued)

© Springer International Publishing Switzerland 2016
P. Linde, *The Hunt for Alien Life*, Astronomers' Universe,
DOI 10.1007/978-3-319-24118-0

(continued)

21. Holborne, Paueans, Galliards, Almains and Other Short Aeirs, "The Fairie Round," performed by David Munrow and the Early Music Consort of London. 1:17
22. Solomon Islands, panpipes, collected by the Solomon Islands Broadcasting Service. 1:12
23. Peru, wedding song, recorded by John Cohen. 0:38
24. China, ch'in, "Flowing Streams," performed by Kuan P'ing-hu. 7:37
25. India, raga, "Jaat Kahan Ho," sung by Surshri Kesar Bai Kerkar. 3:30
26. "Dark Was the Night," written and performed by Blind Willie Johnson. 3:15
27. Beethoven, String Quartet No. 13 in B flat, Opus 130, Cavatina, performed by Budapest String Quartet. 6:37

(NASA)

Appendix C
Tips for Further Reading

General Astronomy and Space Science

ESA	Web site for the European Space Agency	esa.int
ESO	Web site for European Southern Observatory	eso.org/public
HubbleSite	Web site for the Hubble telescope, containing many pictures	hubblesite.org
NASA	Web site for NASA, with space information and images	nasa.gov

Astrobiology

Astrobiology Magazine	Web-based online magazine with latest news about astrobiology	astrobio.net
The Planetary Society	Web site for astrobiology and space research	planetary.org

Solar System

Google Mars	Google's map of the planet Mars	google.com/mars

Exoplanets

Exoplanet Data Explorer	Exoplanet database with emphasis on careful selection of confirmed discoveries	exoplanets.org
The Extrasolar Planets Encyclopaedia	The most complete database of exoplanet data	exoplanet.eu
The Habitable Exoplanets Catalog	Web site with focus on discoveries of potentially habitable planets	phl.upr.edu/projects/habitable-exoplanets-catalog
The Exoplanet Handbook	By Michael Perryman. An expert summary of the scientific front-line in exoplanet research	Cambridge University Press, 2011

(continued)

© Springer International Publishing Switzerland 2016
P. Linde, *The Hunt for Alien Life*, Astronomers' Universe,
DOI 10.1007/978-3-319-24118-0

(continued)

Exoplanets: Finding, Exploring, and Understanding Alien Worlds	By Chris Kitchin An in-depth analysis of exoplanet science	Springer, 2012
SETI		
SETI 2020: A Roadmap for the Search for Extraterrestrial Intelligence	By Paul Shuch (ed.) Future plans for SETI research	Springer, 2010
Confessions of an Alien Hunter: A Scientist's Search for Extraterrestrial Intelligence	By Seth Shostak An account of a SETI expert's search for alien civilisations	National Geographic, 2009
SETI Institute	Web site for the SETI Institute with comprehensive SETI-related information	seti.org
SETILive	The SETI Institute project to involve the general public in SETI signal analysis	setilive.org
The Fermi Paradox		
Rare Earth: Why Complex Life Is Uncommon in the Universe	By P. Ward and D. Brownlee Many arguments to explain why life should be unique to Earth	Copernicus Books, 2000
If the Universe Is Teeming with Aliens... Where Is Everybody?	By Stephen Webb Fifty ways to explain the Fermi paradox	Copernicus Books, 2002
Citizen Science		
Zooniverse	Web site to allow the general public to contribute to scientific research in astronomy and other disciplines	zooniverse.org
Interstellar Spaceflight		
The British Interplanetary Society	Web site for BIS. The society stimulates futuristic projects for interstellar spaceflight	bis-space.com
100 Year Starship	Web site dedicated to the project to launch an interstellar expedition within this century	100yss.org

(continued)

Global risks		
Near Earth Object Program	NASA web site for studies of potentially harmful asteroids	neo.jpl.nasa.gov
Nick Bostrom	Philosophy professor Nick Bostrom's web site about superintelligence, global risks, transhumanism, etc.	nickbostrom.com
The Singularity is Near	By Ray Kurzweil Predictions about future technology advancements	Penguin, 2006

Appendix D
Glossary

Absolute zero The lowest possible temperature −273.15 °C. Atoms in matter are in their lowest energy level.

Absorption lines Dark lines in an otherwise continuous spectrum. Located at specific wavelengths typical for the light that is needed to excite an electron in a basic element. Every basic element generates characteristic lines, like a "finger print".

Accretion disk Rotating disk containing dust and gas moving around a central body like a star or a black hole.

Actuator A mechanical rod whose length can be adjusted by means of applying an electrical current.

Adaptive optics A technique which uses a flexible mirror surface in a telescope to rapidly compensate for turbulence in the atmosphere.

Albedo Measure of reflectivity for a planet. The value 1 corresponds to perfect reflection. The albedo of Earth is 0.3 which is higher than for most planets.

ALMA Atacama Large Millimeter/Submillimeter Array a radio telescope located at an altitude of 5000 m in the Andes, Chile. It contains of 54.12 m disks and 12.7 m disks.

Anaerob A process or an organism that does not need oxygen.

Analysis algorithm A method to extract relevant data from observations using a computer program.

Arcsecond A measure of angle equal to 1/3600 of a degree.

Asteroids Small celestial bodies in the solar system typically fragments from collisions during the formation of the planetary system. Most of them are located between the planets Mars and Jupiter. So far, more than half a million are known.

© Springer International Publishing Switzerland 2016
P. Linde, *The Hunt for Alien Life*, Astronomers' Universe,
DOI 10.1007/978-3-319-24118-0

Astrometry The area of astrophysics that deals with accurate positional determinations of astronomical objects.

Astronomical unit Unit of distance 149.5978707 million kilometres, corresponding to the average distance between the Earth and the Sun.

Ballistic A projectile is ballistic if, after receiving its initial velocity, it follows a trajectory only affected by gravitation and possibly air resistance.

Base pair Bindings that keep together the double helix in the DNA molecule. These normally consist of cytosine adenine, guanine and thymine.

Big Bang The prevailing theory of how the universe was created 13.8 billion years ago. Space and matter were created, and subsequently the universe has continued to expand.

Bioindicator Substance in the atmosphere of an exoplanet which indicates the existence of life. Oxygen ozone and methane gas are examples that can be detected spectroscopically.

Black hole The result of a gravitational collapse for a star which has at least three solar masses remaining when it exhausts its nuclear fuel. A black hole's enormous density and gravitation strongly bends space-time in its vicinity blocking any information inside the event horizon from escaping.

Black-body radiation The radiation emitted from a non-reflecting body only dependent on its temperature. Cool bodies normally radiate at microwave wavelengths, hot bodies at infrared and visual wavelengths.

Blue giants Hot and massive stars with a short and intensive life span. Their masses are several times the mass of the Sun and their high temperatures give them a blue colour.

BOINC system A software platform which enables the general public to participate in scientific investigations using their own home computers.

Bound rotation A tidal effect that influences the rotation of a smaller orbiting body in such a way that it always shows the same hemisphere toward its host star or planet.

Cambrian explosion Geological period about 540 million years ago when life rapidly developed many new forms and species.

CERN European research facility for particle physics located near Geneva, Switzerland. Runs several particle accelerators, among them the Large Hadron Collider (LHC).

Convection Heat transport via the movement of fluid or plasma. The outer layers of the Sun are convective. Red dwarfs are almost completely convective.

Coronagraph Opto-mechanical obscuration of the central field in a telescope allowing the surroundings of the obscured object to be observed. A common application is to create an artificial solar eclipse in order to study the Sun's corona.

Cosmic rays High-energy particle radiation containing extremely energetic particles, mostly protons, accelerated to velocities approaching the speed of light. Originate from the Sun, but also from far-away objects like supernova explosions, black holes and active galaxy nuclei.

Cosmological red-shift Effect showing that distant galaxies are rapidly moving away from us. Interpreted as expansion of space itself. Red shift extends the wavelengths of light, making it redder.

Cryotechnology The discipline in physics that studies the physics of very low temperatures (below $-150\,°C$) and the effects on matter at such temperatures.

Dark energy A repulsive effect which is needed to explain the fact that the expansion rate of the Universe increases with time. The nature of dark energy is unknown but it is estimated that almost ¾ of the total energy content in the universe must consist of dark energy.

Dark matter Unknown form of matter which affects normal matter only through gravitation. For example its effect on the motions of galaxies is quite strong. Its nature is not yet established, but it may be explained by a new type of heavy elementary particle.

Declination Astronomical coordinate analogous to the geographical coordinate latitude, for determination of a celestial body's position on the celestial sphere. $+90°$ correspond to the North celestial pole, while $-90°$ corresponds to the South celestial pole.

Deuterium An unusual isotope of hydrogen in which the atomic nucleus contains one proton and one neutron. Water molecules containing deuterium are called heavy water.

Diffraction limit The highest possible resolution in an optical system that is allowed by the wave nature of light.

DNA Abbreviation for deoxyribonucleic acid, the macro molecule that contains the genetic code in the genome.

Doomsday argument Philosophic argument based on statistical grounds which, among other things, predicts the number of people that will ever live.

Doppler effect Displacement of frequency in light or sound caused by motion of a source relative to the observer.

Dwarf planet An object orbiting the Sun sufficiently large to be round and not a moon to a larger planet. Not capable of sweeping clean its nearby space from other objects. Pluto, Ceres, Eris and Makemake are examples.

Dynamo effect Electromagnetic interaction where electrical currents in a moving conductor generate a magnetic field or vice versa. The effect explains the magnetic fields of planets.

Eccentricity A measure of a celestial body's deviation from a circular orbit. $\varepsilon = 0$ is a circular orbit $\varepsilon = 0.99$ is an extremely eccentric orbit.

Eclipsing variable A double star where the two components lie close together and the orbital plane is seen edge-on from Earth. When the stars in their orbits eclipse each other, their total brightness will decrease, a variation which is very regular.

Electromagnetic wave Coupled electric and magnetic wave phenomena which propagates at the speed of light. We call the waves with the shortest wavelengths gamma radiation, those with the longest correspond to radio radiation. In between are optical light and heat radiation. Electromagnetic waves can also exhibit particle properties and are then called photons.

ESA Abbreviation for European Space Agency consisting of 22 member countries. Since 2011 four East European states have joined the organisation. ESA's Headquarters are in Paris. ESA runs several research facilities and a launch site in French Guyana.

Escape velocity The minimal initial velocity an object must have in order to ballistically escape from the gravitational field of a celestial body. In the case of the Earth it is 11.2 km/s.

ESO Abbreviation for European Southern Observatory, a European research organisation for astronomy, with 16 member states. Headquarters are in Garching, near Munich and in Santiago, Chile. ESO runs three large observing sites in the Atacama desert in the Andes of Chile.

ETI Abbreviation for Extraterrestrial Intelligence sometimes used as synonym for extraterrestrial alien.

ETI-signal A signal transmitted from a hypothetical extraterrestrial intelligent civilisation.

Eucaryotes Organisms containing cells with a nucleus surrounded by a membrane. All higher organisms, including man, belong to the eukaryote group.

Extremophile Organism that can survive extreme conditions for example cold, dry, salty or hot environments.

FFT Abbreviation for Fast Fourier Transform, a mathematical computer algorithm which enables easy identification of periodic signals (frequencies) in, for example, a radio signal.

Galactic magnetic field The total magnetic field of the Milky Way. Remotely generated cosmic rays with electric charge, are deflected in different directions by this field.

Gamma-ray burst Extremely powerful burst of gamma-rays observed from distant galaxies. Possible explanations include supernova explosions or collisions between two neutron stars.

Gemini observatory American observatory with two 8.2 m telescopes one located in Chile, the other in Hawaii.

Genome Designation for the total genetic information coded in the DNA molecule.

Globules Small spherical bodies. In astronomy the term is used for collapsing volumes of dust and gas, an early stage in the formation of stars.

Gravitational microlensing Method of discovering exoplanets by measuring the light amplification which arises when two stars are aligned with each other. The gravitation of the foreground star bends space-time so that it works as a lens for the light of the background star.

Gravitational resonance Coupling caused by gravitation between two celestial bodies such that their orbital and rotational periods are interlocked. 1:1 resonance corresponds to bound rotation, but there are other couplings, expressed by integer number ratios.

Gravity assist Method to change (usually accelerate) the speed of a spacecraft by letting it pass close to a planet. The planet's own motion can, through a slingshot effect, act on the spacecraft.

Habitable zone The zone around a star in which a planet can exist with liquid surface water. In this zone the heat from the star is just sufficient to avoid both boiling and freezing of water.

Heat capacity A material's capacity for storing heat. Solid substances have higher heat capacity than for example, gases.

Hubble Space Telescope (HST) An optical space-based telescope with a 2.4 m mirror. Starting in 1993 it has taken pictures with unprecedented detail, resulting in advancements of many areas of astronomy. Named after American astronomer Edwin Hubble.

Index of refraction Optical characteristic of transparent materials that specifies how light is refracted at oblique entrance.

Infrared radiation The part of the electromagnetic spectrum with longer wavelengths than optical light and which we perceive as heat.

International Astronomical Union (IAU) Worldwide organisation for professional astronomers. Holds a General Assembly every third year. Officially designates names for stars and planets establishes standards, etc.

Interplanetary scintillation Random variations in the intensities of radio sources whose signals pass through the solar wind before reaching Earth. Analogous with the atmospheric turbulence effect which causes the apparent naked-eye twinkling of stars.

Ionising radiation Radiation with short wavelengths (ultraviolet X-ray, etc) with sufficient energy to remove electrons from an atom. This creates an imbalance, causing the atom to have a positive electric charge. It is then called an ion.

Keck telescopes Two American giant telescopes located at an altitude of 4000 m in Hawaii. Both have segmented mirrors with a diameter of 10 m. They have been financed from the W M Keck foundation.

Kepler project NASAs ongoing study using the Kepler satellite to search hundreds of thousands of stars looking for exoplanets. Using the transit method, almost 2000 exoplanets have been found so far.

Kepler's first law States that planetary orbits are ellipses with the Sun in one of the two foci. Published by Johannes Kepler in 1609 based on Tycho Brahe's observations of Mars.

Kepler's second law A line joining a planet and the Sun sweeps out equal areas during equal intervals of time. Published by Johannes Kepler in 1609.

Kepler's third law The square of the orbital period of a planet is directly proportional to the cube of the semi-major axis of its orbit. Published by Johannes Kepler in 1619.

Lagrange point Stable point in an orbit related to two larger celestial bodies, for example the Sun and the Earth. There are five such points in every two-body system. They are gravitationally stable regions, suitable for locating satellites.

Late heavy bombardment A period about 4 billion years ago when a large number of larger bodies collided with the inner planets. Possibly caused by gravitational interference from Jupiter and Saturn.

LOFAR Abbreviation for LOw Frequency Array a giant radio telescope, optimised for long radio wavelengths. Contains about 20,000 simple antennae, located in groups in several European countries.

Luna Soviet programme for exploration of the Moon. More than 40 launches were made 1958–1976, 15 were successful. Among other things, Luna 3 took the first pictures of the far side of the Moon. In 1970, Luna 16 returned samples of lunar soil to Earth.

Magellanic clouds, Large and Small The two satellite galaxies of the Milky Way at a distance of 160,000 and 200,000 light years, respectively. In the Southern sky they appear to the naked eye as two diffuse clouds of light.

Magnitude The measure of a star's brightness. The scale is inverse—the larger the value the fainter the brightness. A magnitude difference of one magnitude corresponds to an intensity difference of 2.512 times. Thus, five magnitudes exactly correspond to a factor 100 times intensity difference.

Main sequence The banded zone in the Herzsprung-Russell diagram where the stars spend most of their life span in a stable mode. Our Sun has currently spent about 5 billion years on the main sequence.

Metallicity Is a measure in astronomy of a star's content of "metals" i.e. elements that are not hydrogen and helium. These elements were created during the Big Bang, while the others are almost exclusively created in nuclear reactions inside a star.

Microbe Same as micro-organism, an organism, that is too small to be seen by naked eye.

Micrometre Unit of length same as a millionth of a metre, i.e. a thousandth of a millimetre. Also known as micron.

Milliarcsecond A thousandth of an arcsecond. A small two metre wide crater on the Moon subtends an angle of 1 milliarcsecond as seen from Earth.

Molecule A system of interconnected atoms. The connection between atoms works with different electric bonds. As an example water consists of two hydrogen atoms and an oxygen atom.

Moore's law Rule of thumb that predicts that the computing capacity of computers doubles about every 18th month. Has approximately been correct over the last 40 years.

Nanometre Unit of length, a billionth of a metre. Often used to describe distances on an atomic level. The diameter of a helium atom is about 0.1 nm, the wavelength of yellow light is approximately 550 nm.

Narrow-band signal Radio signal that contains a single or small number of frequencies. For visual light the corresponding concept would be a single colour (for instance light from a laser).

Negative coupling A concept in control theory which means that a system can dampen itself and remain stable. A thermostat is an example.

Neutrino pulses A hypothetical method to communicate using neutrinos. Such elementary particles are very common but extremely difficult to detect.

Nuclear winter As a result of a nuclear war, enormous quantities of dust would be thrown into the atmosphere, reduce the solar input and cause a colder climate during a prolonged period of time.

Organic molecules Molecules that are based on various forms of carbon compounds. Examples of simple hydrocarbons are the gases methane and ethane.

Parsec A unit of length. The distance from which the Earth's distance to the Sun (1 AU) is seen at an angle of one arcsecond. Corresponds to 3.26 light years.

Percolation A filtering process in which a liquid passes through a porous medium (cf. a coffee maker). Can be simulated mathematically in order to study various physical processes.

Photometry Accurate measurement of the brightness of a celestial body. Filters with different colours (wavelength bands) are used to define different magnitudes. Typical detectors are CCDs photo multipliers and photographic film (obsolete).

Photomultiplier Very sensitive light detector, in the form of a vacuum tube that electronically amplifies incoming light by about 100 million times. Although not a panoramic detector it is still used, for example, in photon counting applications.

Plate tectonics Theory which explains the continental displacements as a result of large plates moving on top of a viscous hot magma inside Earth.

Polarisation Light can be perceived as an electromagnetic wave. The orientation of the electrical field corresponds to its polarisation. Ordinary light is a combination of all angles while oblique reflection on a reflective surface will result in linear polarisation.

Prime number Is an integer number larger than 1 and only divisible by 1 and itself. The first are 2, 3, 5, 7, 11 and 13. Sometimes proposed as part of an interstellar message, revealing a non-natural origin.

Proper motion The part of a star's space velocity which is directed sideways. It is usually measured in arcseconds per year.

Quantum mechanics Theory in modern physics describing the characteristics of energy and matter in micro-cosmos, for example the discrete energy levels in an atom. The Heisenberg uncertainty principle and the particle/wave dualism are important concepts.

Quasars Objects from the early period of the universe having an enormous luminosity, sometimes equal to hundreds of galaxies. Matter's interaction with enormous black holes causes the energy dissipation. Most quasars are more than 3 billion light years away.

Radial velocity A star's real velocity along the line of sight. Can be measured using the Doppler effect.

Radian Measure of angle defined from the unit circle. $1 \, \text{rad} = 360/\pi$ degrees $\approx 57.3°$.

Red dwarf Star with a mass between a half and a tenth of a solar mass. The surface temperature is less than 4000 °C which gives it a red colour. The active life span can be hundreds of billions of years.

Red giant Star at the end of its development. Becoming unstable it expands to perhaps a hundred times its original size. Simultaneously, the surface temperature decreases, giving it a red colour. Its total luminosity increases due to the increased total surface.

Right ascension Astronomical coordinate analogous to the geographical coordinate longitude, for determination of a celestial body's position on the celestial sphere. Measured in hours along the celestial equator toward the East. Its origin (0 h) is the vernal equinox. A complete circle corresponds to 24 h.

Retrograde Used to describe for example, the motion of a planet which is orbiting or rotating in opposite direction, compared to the dominating direction of the system to which it belongs.

RNA Abbreviation of ribonucleic acid, a single helix macro molecule, which today has the function, among other things, of speeding up chemical processes in the cell.

Schmidt telescope A special optical design for a telescope which allows for sharp images over a large field of view. Originally developed by Bernhard Schmidt in 1930.

Seismic vibrations Stars generate mechanical vibrations in a way similar to a tuning fork. These can be observed utilising the Doppler effect. From this information properties inside the stars can be deduced.

SKA Abbreviation for Square Kilometer Array a gigantic future radio telescope, that is planned to consist of 3000 disks, each with a diameter of 15 m.

Snow line The distance from the host star beyond which, during the creation of a planetary system, the temperature is low enough to enable water to form ice. Inside the snow line, water exists only in vapour form, subject to pressures from the stellar output radiation.

Space-time A concept describing a four-dimensional space where the three spatial dimensions and time are interconnected. Matter affects the geometry of space-time, which is mathematically described by Einstein's general theory of relativity.

Spectrograph Instrument for detailed studies of the wavelengths of light, especially absorption and emission lines. Also known as a spectrometer.

Spectrum The separation of light into its wavelengths or colour components. The rainbow is an example of a spectrum.

Spitzer telescope Space telescope for studies of the infrared region of light. Launched in 2003. 2009 the cooling medium was used up and nowadays only a couple of the instruments are in use. Named after American astrophysicist Lyman Spitzer Jr.

Sublimation A process during which a substance is converted from a solid to a gas without any intervening liquid stage.

Time-resolved observations Observations made at high speed in order to study fast processes. An example is high-speed photography.

Transhumanism An international movement advocating usage of advanced technologies to enhance and increase the physical and mental capabilities of humans.

Vacuum energy A theoretical concept originating from quantum mechanics. The basic idea is that empty space contains energy in the form of continuously created and annihilated virtual particles. Some phenomena, for instance the Casimir effect, seem to verify the validity of the concept.

Very large telescope (VLT) Europe's—and one of world's—largest telescopes. Located on a mountain top in Northern Chile at an altitude of 2600 m. It is operated by ESO and contains four 8.2 m telescopes.

White dwarf A star remnant with very high density, approximately the size of Earth. It is formed as a result of a gravitational collapse of a star having less than 1.4 solar masses remaining after the nuclear fuel is exhausted.

Index

© Springer International Publishing Switzerland 2016
P. Linde, *The Hunt for Alien Life*, Astronomers' Universe,
DOI 10.1007/978-3-319-24118-0